The Burden of Choice

The Burden of Choice

•••••••••••••••••••••••••••••••••

Recommendations, Subversion, and Algorithmic Culture

JONATHAN COHN

Rutgers University Press

New Brunswick, Camden, and Newark, New Jersey, and London

Library of Congress Cataloging-in-Publication Data

Names: Cohn, Jonathan, author.
Title: The burden of choice : recommendations, subversion, and algorithmic culture /
    by Jonathan Cohn.
Description: New Brunswick : Rutgers University Press [2019] | Includes bibliographical
    references and index.
Identifiers: LCCN 2018031140 | ISBN 9780813597829 (cloth) | ISBN 9780813597812 (pbk.)
Subjects: LCSH: Consumer behavior. | Consumer profiling. | Consumption (Economics)
Classification: LCC HF5415.32 .C643 2019 | DDC 306.3—dc23
LC record available at https://lccn.loc.gov/2018031140

A British Cataloging-in-Publication record for this book is available from the British Library.

♾ The paper used in this publication meets the requirements of the American National
Standard for Information Sciences—Permanence of Paper for Printed Library Materials,
ANSI Z39.48-1992.

www.rutgersuniversitypress.org

Manufactured in the United States of America

For Jaimie

Mike Pesca ✔
@pescami

I think you may be confusing freedom with Amazon Prime.

> **Paul Ryan** ✔ @PRyan
>
> Freedom is the ability to buy what you want to fit what you need. Obamacare is Washington telling you what to buy regardless of your needs.

6:20 PM - 22 Feb 2017

Frontispiece: Tweet from Mike Pesca in response to Paul Ryan, February 22, 2017

Everything is a recommendation.
—Xavier Amatriain and Justin Basilico, Netflix

Contents

The Burden of Choice

Introduction

•••••••••••••••••••••

Data Fields of Dreams

> For a long time, it seemed that, even in a machine age, certain things always would require a purely personal performance. Things like picking out Christmas presents or perhaps teaching football plays. But automation marches on, with the result that the human effort has been practically eliminated from even those ventures by machines that tell everything from how to block on a quarterback sneak to how to please the girlfriend at Christmas.

In the 1950s and 1960s, the Neiman Marcus[1] department store in Dallas, Texas, became world renowned for its opulent displays and extravagant marketing campaigns. Its Christmas catalogs featured ultraluxurious fantasy items that included, over the years, a live Black Angus bull with a sterling silver barbecue cart, a gold-plated toilet seat, a baby elephant, Napoleon's glasses, a jetpack, and Noah's ark—complete with all endangered species. In addition, they also often advertised matching gift sets, including "his and hers" airplanes, dirigibles, submarines, robots, and ancient sarcophagi. According to apocryphal legend, this catalog was at one time the most stolen piece of mail in the United States. Along with its Christmas catalogs, Neiman Marcus was also famous for its Fortnight festival, a gala shopping experience that celebrated the products and culture of a different country each year in late October. The first Fortnight in 1957 showcased French culture with massive displays of exotic goods, fashion shows, a gala ball, and a Henri de Toulouse-Lautrec art exhibition, along with appearances by many ambassadors, foreign dignitaries, and celebrities.

Within this context of fantastical consumerism and seemingly unlimited choice, IBM showcased the pleasures and promises of both the universal

1

computer and the automated and algorithmically generated recommendation for the modern bourgeois American shopper. During the 1961 Fortnight— which celebrated American rather than foreign products for the first time— Neiman Marcus featured an IBM 1401 computer Gift Advisory System that recommended personalized gifts based on answers to customer questionnaires.[2] These multiple-choice forms asked patrons about the age, sex, vocation, hobbies, habits, and marital status of the person for whom they were buying the present—along with the customer's price range.[3] Using punched cards, the computer would then search through a database stored on magnetic tape of "2,200 to 2,800 Neiman Marcus items in its memory and pick 10 that seem[ed] best to fit the individual described."[4] According to *Businessweek*, IBM expected to process somewhere between twenty thousand and forty thousand punch cards during the Fortnight. Customers reported receiving a wide range of gift suggestions, including gold soap dishes, velvet jackets, espresso makers, and wristwatches. For one guest pretending to be President John F. Kennedy in search of a gift for his wife, the computer suggested a yacht would be appropriate.[5] With similar gusto, a "wise guy asked the machine to select a gift for a six-year-old child-bride who had eleven children. Unruffled, the machine printed: 'You have described a very unusual person. If you will produce the person, we will produce a gift.'"[6] Journalists reported that these automated recommendations turned Christmas shopping from a "burden, a chore, [and] a tiresome bother" into a fun, "simple, speedy service."[7]

Such technologies and values fit comfortably with our current neoliberal (or simply capitalist) fascination with generating efficiency and simplicity in all areas of life through automation. Rather than celebrate the joys of strolling through aisles and enjoying the variety and vastness of America's bounty and consumer choices, Neiman Marcus instead relied on these automated recommendations to make the shopping experience both more fun and more productive. Through these recommendations, making a choice was framed as a "burden," while automated computer technologies became the solution. These recommendation programs taught the bourgeoisie to treat their privileges and options as a burden they could pleasurably cast off to technologies and technocrats.

Within the context of this Fortnight, which traditionally featured only foreign products, the futuristic and extremely gimmicky Gift Advisory System worked simultaneously to exoticize American goods and domesticate computers—technology that was at that time associated primarily with military and industrial applications like the "periodic compilation and transmission of stock market prices, the guidance of a blade cutting steel along a straight line and the accurate transmission of a subtle color blend [to an electronic monitor]."[8] The IBM 1401 was marketed not as the best or biggest computer but rather as the cheapest and most practical machine for "ordinary businesses that simply wanted to run their finance and accounting systems more efficiently, and in particular, to replace single-purpose electromechanical machines that used levers, knobs, and plugboards to program their operations."[9] The 1401s were smaller, cheaper, and easier-to-use computers meant to represent a new paradigm in which business, automation, and data management became inseparable processes. Rather than stress the futuristic aspects of the 1401 with connotations of artificial intelligence and mechanistic agency, IBM toned down this rhetoric by focusing on its humble practicality and affordability. Neiman Marcus's recommendations became the commercial epitome of IBM's efforts to cast computers as servants designed to help the bourgeoisie manage their lives.

The IBM 1401 was the first widely available "universal machine" capable of doing more than one task at a time, and it could therefore be advertised to a wide variety of businesses through public showings like the Fortnight. For many Neiman Marcus customers, this may have been their first physical interaction with a computer; thus, this event served as a vivid demonstration of how these machines could affect their social lives and relationships. If a computer could know "how to please the girlfriend at Christmas," what other "purely personal" tasks could it perform for us? This marketing was effective: by 1965, half of all computers in the world were some type of the 1401.

Long before Amazon, Netflix, and others popularized such recommendations, the Fortnight emphasized how recommendations—and the computers that created them—were, even then, weaving together the global economy and the domestic sphere; one IBM executive asserted, "We don't want to get the reputation of having automated Santa Claus, but the truth is that our computers are the best helpers he's ever had since he was smart enough to hire reindeer."[10] Here, Santa becomes a precursor and personification of what we now call big data: he has a miraculous ability to surveil "everyone" at all times; he then uses a simplistic (if generous) algorithm to divide the nice from the naughty, and he doles out personalized rewards and punishments accordingly. Or, rather, this is the softer and kinder vision of big data presented to affluent white America, where one most often comes into contact with it through the seemingly benign worlds of online shopping and social networking. For many others across the globe, big data more often appears as Krampus, marginalizing populations wholesale and facilitating drone strikes.[11] Yet, even here in this most charitable of representations, there are hints of anxiety stirred up by IBM's ability to "automate" Santa and his many elves: If a machine could automate Santa or replace an army of clerical elves, what then of the role of the American retail worker or the discerning customer? And how did the affluent shopper's excitement over these technologies enmesh him or her in a massive economic transformation?

Although IBM's Gift Advisory System was marketed in terms of its usefulness and efficiency, it also illustrates the many ways in which our experiences
with big data—and recommendation technologies more specifically—have long
been gendered, classed, and racialized. In a promotional photo taken by IBM
to advertise its computer as a tool for personal advice, a clearly affluent blond
white woman with a fur hat stands in the foreground as she holds her leather
gloves and gift suggestions proudly while a tight-lipped white male IBM worker
leers at her. Above his head is a sign pointing at him that proclaims the fast
speed of the IBM 1403 printer. This photo illustrates the nexus of relationships
that automated recommendations generate. In the background stands the
actual computer, complete with Neiman Marcus's product database, stored on
tape reels. A series of impediments blocks the woman and the rest of the public from directly accessing, manipulating, creating, or destroying these data. As
a consumer, she is not allowed to transform or touch the database itself. Given
that the woman's interface with the computer at this point is through the printer
that blocks her way, she cannot affect the actual range of her possible shopping choices. Her only decision is whether to take the computer's advice.

Moreover, through the position of the man and woman and the direction
of their gazes, the photo depicts the power dynamics implicit in the woman's
choice. The man's position above the woman and in the midst of huge pieces
of hardware illustrates that although the computer generated the recommendations, it did so only because he or someone like him programmed it for this
special purpose. While the woman is in focus and takes up most of the space,
the recommendation and man are at the center. The arrow, pointing at the
man; the man's gaze, directed at the woman; and her gaze, directed to the
side at what one might imagine to be a cheering crowd, illustrate a flow of not
just advice but also agency. To complicate matters further, while blurry, the gift
advice is not for the woman but rather for her husband; the page recommends
that she buy him a gold watch, cuff links, a desk set, a gold pen and pencil set,
and/or a tailored suit. This advice places her in the position of a dutiful wife
who shops in order to literally make her white-collar husband look good. Thus,
this photo depicts a complex, flowing hierarchy of authority and agency in
which the computer and the white, patriarchal order are the ultimate arbiters.
In this system, the computer generates the recommendations and the recommendations help replicate the already established social order. While in this
case the values promoted are clearly those of white bourgeois America, such
systems could at least theoretically be used to reify most any identity category, for better or, more likely, worse.

IBM's Gift Advisory System is just one of many examples of how automated recommendations have reflected and amplified our cultural desires
and beliefs since nearly the beginning of the computer era. In this book I
examine how recommendations—and particularly algorithmically generated

FIG. I.1 Neiman Marcus Gift Advisory System, IBM Archives, 1961

recommendations—are produced and now experienced on the World Wide Web. In particular, I focus on the cultural anxieties and political debates that engendered these recommendations and that they themselves continue to produce. I focus on technologies that automatically recommend things— products, entertainment, news, romantic partners, and even cosmetic surgery operations—to users in online spaces, as well as the industries that depend on these technologies. These industries all rely on what are widely known as digital recommendation systems, which are derived from an analysis of aggregated user data to lead people toward certain objects and away from others. Today, these automated recommendations help us navigate the web, and the technologies they are built on undergird the digital economy through the personalized advertisements that they afford. Throughout this book, I use a wide range of examples, including social networks, digital media distribution companies, online dating sites, and cosmetic surgery practices, to trace how the creation and giving of personal advice has always been a key facet of computing and how recommendations became a central trope of digital culture.

Like many during this moment when #BlackLives rarely seem to matter and when reports suggest that Donald Trump's presidential campaign used recommendation algorithms and similar technologies not only to increase his votes but also to suppress Hillary Rodham Clinton's via "fake news," I am fundamentally concerned with how the recommendation has come to serve as a form of ideological control that facilitates patterns of exploitation and inequity.[12] A study of the history and current use of recommendations necessarily elucidates what those in power value and how those values flow—nationally and globally—through discourses, technologies, cultures, and relationships.

These recommendations and the technologies that underpin them too often attempt to reify the identities of users according to harmful stereotypes of race, class, gender, sexuality, and other markers of identity. The examples are legion: Google and Facebook are more likely to present recommendation-system-generated ads for criminal-related services such as background checks and bail bonds to users with stereotypically African American names than to those with typical Caucasian names.[13] Google has also been criticized for being less likely to show women ads for high-paid jobs. Likewise, it is notorious for autocompleting search queries with racist, sexist, and homophobic phrases and questions.[14] During the 2016 U.S. presidential election, YouTube's recommended videos, regardless of search terms, were 80 percent more likely to be favorable to Donald Trump.[15] At the same time, YouTube's Restricted Mode for educational institutions censored LGBTQ content as "unacceptable for kids."[16] And while Amazon does not allow people to sell pornography on its site, many have criticized it for not just selling but also recommending Nazi paraphernalia and white supremacist literature to those who have already presented an interest in similar items.[17] Together, these recommendations and the practices that surround them privilege white heteromasculinity as not just the dominant but also the best identity.

While these technologies could theoretically be designed to encourage the creation of a more equitable world, I have not yet found any that do. This is a result, in part, of how these technologies determine what we desire as individuals rather than what we need as a collective. They therefore work to expand rather than end capitalist forms of exploitation, domination, and oppression. Until we as a world actually start to want equality—as more than an empty rhetorical gesture—and not in spite of but rather because of all the perceived difficulties that it portends, our technologies will never recommend we take the laborious steps that are necessary. In the meantime, they facilitate what Gilles Deleuze called the modern control society. In an effort to understand how capitalist exploitation continues unchecked, Deleuze argued in the 1980s that contemporary society is governed by a logic of control that appears to value flexibility, immateriality, individuality, and freedom, but only insofar as it also encourages the continued "exploitation and the degradation of material environments on a global scale."[18] As many have since argued, control is fundamentally connected to digital technologies; for instance, Seb Franklin argues that this control "both defines and instrumentalizes individual actors and groups, whose conditions of social existence are now premised on statistical predictive models and decisional states that rest on a conceptual as well as a technical digitization of the world."[19] While digital industries continually sell their technologies as tools of global empowerment and democratization, their exploitation of users, workers, and the environment continues unchecked.

Recommendation systems are a primary technology of this control society. Just as control more generally works by presenting exploitation as freedom, recommendation systems privilege the "free choice" of users as a synecdoche of their unique individuality, self-worth, and authenticity while, in fact, always guiding users toward certain choices over others in order to encourage them to better fit in with those the system recognizes as being like them. Just as control more generally divides people in extremely specific ways that do not necessarily follow distinctions along lines of race, class, gender, sexuality, nationality, and so on (though they often do), recommendation systems present consumer desires, preferences, and tastes as key to divining who you are, where you fit in, and how to become the best you that you can be (or at least the best capitalist subject you can be).

Even as we use these technologies to discover what we might most desire, they work to define and shape our desires, our decisions, and ourselves through interpellative processes. Like Deleuze, Louis Althusser criticized the ways in which capitalist society continues to reproduce the conditions of its production and stabilizes the inequitable relationship between the proletariat and bourgeoisie. He illustrated how the state educates the proletariat through schools, churches, and other ideological state apparatuses to accept their subjectification and the bourgeoisie to deliver it. Through these institutions, he argued, we are continually interpellated as ideological subjects by those around us, and we show our acceptance of this subjectification even by simply responding to someone yelling at us from across a street. While we typically think of such a response as an act that illustrates our free will and agency as individuals, Althusser argued that it actually proves we understand the customs and etiquette of our society and are willing to obey them; thus, we continually, in every tiny action, show that we accept our subjectification.

In direct contrast to this discourse, companies typically advertise their digital technologies as empowering tools that allow us to make more and better choices as individuals. Yet Mark Poster, Lev Manovich, and many others have argued that digital technologies (especially the database) continue to interpellate users in often harmful ways.[20] Manovich states that while learning to use a computer program—whether learning what all the buttons do or how to most efficiently complete a task—we are also learning how the program wants us to act. Updating Althusser, Manovich states that interactive media ask us "to mistake the structure of somebody's else [sic] mind for our own."[21] In the process, interactive media teach us to identify with its creators' mental structures—and values.

Interpellation and control are concomitant (and often indistinguishable) forces in that they both work by paradoxically presenting subjectification and dependency as symbols of autonomy and freedom. Recommendation systems interpellate us as subjects by asking us to continually consider not just whether we want what they offer but also why they imagine that we are the kind of subjects that could conceivably desire their options. When we try to make our recommendations "better" by clicking on things we do like and ignoring everything else, we may imagine that we are teaching the program, but it is also teaching us; we feel as if we are mastering the program, but all the while, the program is mastering us. Generated recommendations teach us not just what we might like but also what similar subjects like—and suggest that we should like these things, too. In the process, these recommendations help reproduce our inequitable society by teaching us how to better fit into our established place within it. This focus on individual choice instantiates and automates Althusser's belief that "post-industrial capitalism's espousal of the ideology of choice is not a coincidence but rather enables it to perpetuate its dominance."[22] Ultimately, this amplification of desires only reproduces the devastating effects of modern capitalism by reinforcing our placement within it.

Yet the examples I discuss in the following chapters illustrate that in practice, the controlling, interpellative effects of recommendation systems are far from total; they are as susceptible to misunderstandings, misuses, subversion, and sabotage as any other tool. During this period of increasing inequity, environmental decay, and global unrest, it may be tempting to believe that these technologies manipulate us in ways we are powerless to stop. The precarious nature of the world may then cause many to believe they are not responsible for their actions. However, this belief also obfuscates all the various ways we continually critique, ignore, laugh at, negotiate with, and otherwise respond to these recommendations not as benign or foregone conclusions but rather as an oppositional and opposable force. It ignores all the agency we do still have and only makes it harder to recognize how we might do more than just accept the recommendations given us.

FIG. I.2 Screenshot of *The Babadook* Netflix recommendation and use at Gay Pride Parade

For instance, users have often appropriated problematic recommendations for their own uses. When one user on Tumblr posted an image of their Netflix LGBT movie recommendations, which mysteriously included the documentary *Room 237* (2013) and the horror film *The Babadook* (2015), Twitter and Tumblr users turned this image into a meme and the Babadook became a symbol of gay pride ("The B in LGBT stands for Babadook"; #babashook).[23] While scholars like Ted Striphas, Alexander R. Galloway, Ed Finn, Frank Pasquale, and John Cheney-Lippold tend to focus on examples in which algorithms get it right and illustrate their omnipotent and omnipresent potential to know us better than we know ourselves, our interactions with algorithms may also lead to surreal errors and unpredictable responses. The Babadook example highlights how, like every other technology and representation, recommendation systems can take on a life of their own. Yet in making this meme, its creators seemed to avoid a deeper critique of Netflix's recommendations, as did those who spread it. At best, this was a careless error that illustrates how little attention Netflix was paying to LGBT content, and at worse it conflates horror with homosexuality. The fact that everyone assumes this was a simple error and that it did not result in any anger toward Netflix demonstrates how algorithms seemingly separate corporations from their more problematic actions; in turn, corporations mobilize this mistaken belief in order to divorce themselves from their own carelessness and bias. In the following chapters I focus on instances in which this opposition became particularly clear in order to illustrate not only how existing conceptions of gender, sexuality, race, and class shape what these technologies recommend but also how one might resist these recommendations.

While there is a growing body of scholarship on the importance of algorithms today and the various industries that rely on them, this scholarship tends to be either overly dystopian or utopian in tone; some frame automated recommendations as a great democratizing innovation, while others suggest they are fascistic and oppressive.[24] Together, these works tend to fetishize algorithms and treat them as entirely new innovations of our contemporary digital landscape. They also portray typical users as willing to unthinkingly take whatever recommendations are offered them. The central questions in many of these studies are whether we can and should trust companies with our data and whether we should trust their algorithms to supply us with an accurate depiction of ourselves and the greater world. (The answer to both questions, of course, is *no*.) To some extent, the question of whether algorithms are good or bad actively disregards acts like queering Babadook and all of the other creative, pleasurable, and oppositional ways users can react to and repurpose algorithmic output. These acts are typically viewed as trivial and silly rather than as seeds of algorithmic dissidence, as I argue they are. Here, I am interested in how these moments reveal the limits of the recommenders' reach and the methods by which users at times push back. With an eye to the deeper historical contexts within which contemporary recommendations are made, I explore the complexity of such recommendations and how we use and misuse them.

While my focus here is limited to tools that specifically rely on digital recommendation systems or refer to their services as recommendations, these systems and the logic of recommendations are everywhere; both their potential and their implications are wide ranging. Even Yahoo, one of the earliest and most successful search engines, began in 1994 as a relatively short list of recommended websites called "Jerry and David's Guide to the World Wide Web." While much has changed since then, this recommendational logic and the technologies that support it are still a core element of all search engines. Efforts to define the differences in the ways in which search engines, recommendation systems, and personalized advertisements work highlight how extensively these activities now overlap.[25] Indeed, it is virtually impossible to make any decision online (and sometimes offline) without encountering recommendation systems. They help us pick shoes on Zappos.com, they help us decide which link on Google to click on, and they even calculate bail amounts in courtrooms across North America. At the same time, these algorithmic technologies blur the line between benign helpfulness and manipulation. As the case of the political consulting firm Cambridge Analytica, which covertly used private Facebook data to generate duplicitous Trump ads, made very clear, it is now often impossible to tell what decision-making help is actually a product of paid advertising and what is "impartial."[26] Perhaps the most scandalous aspect of this episode was that Cambridge Analytica's actions are not atypical but rather a quotidian aspect of online advertising. And indeed, while my focus is on how the recommendation functions within digital culture and technologies, many of my findings are relevant to other ways in which we use technologies to make decisions.

While recommendations are now a largely ubiquitous part of the web, in this book I focus on the period starting in the mid-1990s, when digital recommendations were novel. In conversation with Brian Winston's and Lisa Gitelman's critiques of the "newness" of new media, this choice stems from the assumption that "looking into the novelty years, transitional states, and identity crises of different media stands to tell us much, both about the course of media history and about the broad conditions by which media and communications are and have been shaped."[27] Furthermore, during these transitional periods, the cultural politics and social anxieties about new technologies often become visible. Throughout this book, I examine various ruptures in the history of automated recommendations when such cultural meanings and anxieties boiled to the surface in the form of controversies, debates, popular news stories, and other mediated representations. I do at times discuss more recent examples in order to illustrate how earlier debates about recommendations persist, but the main focus of this book is on this earlier historical moment. Companies like Google, Amazon, and Netflix now present their recommendation technologies as neutral, objective, benevolent, mathematical, and natural. Primarily through a discursive analysis focused on how producers, users, and journalists described and discussed recommendation technologies during this early period, I will show that automated recommendations (and algorithms more generally) are not natural, neutral, or benevolent technologies but rather shape—and have been shaped by—changing conceptions of gender, sexuality, race, and class.

Corporations paradoxically present these technologies simultaneously as revolutionary products of our digital era and as timeless practices as ancient as humanity itself. While I focus on the period when recommendation technologies were "new," I continually address how, in a sense, they are also always already old. These technologies have histories, and their current potential and limitations are a result of the perceived needs, desires, and fears of industries, consumers, and governments. Even the "recommendation" itself, an activity that humans have perhaps always engaged in, has a history. As Gitelman and Geoffrey B. Pingree state, such histories allow us to see "how interpretive communities are built or destroyed, how normative epistemologies emerge."[28] While it is easy to imagine that our current fixation on issues such as privacy, surveillance, and the changing nature of personal relationships is brought on primarily by digital technologies, a discussion of the longer and intertwined history of recommendations and predigital computer technologies illustrates how these topics have long been woven into the fabric of our daily lives. Indeed, these concerns helped make digital culture what it is.

At the same time, figuring out how to study and critique algorithmic appa-ratuses is notoriously difficult. The algorithms and data that undergird com-panies like Google, Netflix, and Amazon are ostensibly their most important assets and what sets them apart from their rivals; thus, the specific code and statistical models used by such commercial websites to generate recommenda-tions are largely trade secrets and private property hidden from the prying eyes of those who would either copy or critique them. Beyond what companies vol-untarily report, and what we can gather from our personal experiences and edu-cated guesses, we rarely have any idea of how they use, store, and process our information. Pasquale argues that this secrecy is the most disturbing and harm-ful aspect of algorithms, especially as corporations tend to use this secrecy to generate profits and hide unscrupulous acts.[29] He is certainly right, though there is no guarantee that making algorithms more perceptible or subjecting them to more oversight would necessarily make them less damaging. Consid-ering this systematic obfuscation, then, *representations* of these algorithms and their effects play a crucial role in revealing how we use, experience, and think of them—if we think of them at all.

This, then, is a book about the ways scholars, journalists, and marketers represent algorithms and their histories; it is also a book about how these representations shape the way we experience and use said algorithms. It does not pretend to be a definitive or complete history of the recommendation, or even the automated recommendation, as the scope of such a project is far too vast. Yet I do consider throughout how our uses for, relationships with, and experiences of algorithms and recommendations are not natural or inevita-ble but rather have been shaped by various people, companies, organizations, and industries. By locating how things could have gone differently, we can begin to imagine how we might change them ourselves to better reflect the values of equity and social relations many of us hold. Through close readings, historical research, and qualitative analysis, I model one particularly promising avenue for the study of algorithms and their effects on our culture.

Algorithmic Culture

While computer scientists typically study algorithms to determine whether they accurately and efficiently accomplish the task they are designed for, since the mid-2000s, more and more humanities scholars have begun considering the larger role algorithms play in shaping our culture, economy, politics, and every-day reality. Thanks to the work of a bevy of inspiring scholars, the question is no longer whether seemingly innocuous algorithms can spread values and affect culture but rather how and to what extent they do.

But algorithms are still mysterious and largely invisible things; I will start with the basic question, What are they and how do they work? Janet H. Murray broadly defines algorithms simply as "a sequence of steps that need to be followed to accomplish a task."[30] Yet as Tarleton Gillespie points out, while software engineers may think of algorithms as "quite simple things," the broader public imagines that they are "unattainably complex."[31] Both engineers and the public also associate algorithms with a mathematical rationality and objectivity that makes those algorithms appear to be fair and incorruptible.

Examining the actual effects and outcomes of algorithmic calculation can reveal that what was intended by the original coder may have a variety of unforeseen consequences. Like Gillespie, I use the term *algorithm* as a synecdoche of the larger apparatus within which it is entangled. In addition to the algorithm itself, this apparatus includes the "model, target goal, data, training data, application, hardware," and its "broader social endeavor."[32] Simply analyzing an algorithm without taking into account the data it works on and how companies and users employ them is like analyzing a piece of literature outside its cultural context: you may be able to make an argument concerning its formal qualities, but it runs the risk of being ahistorical and can lead to a false understanding of how and why the algorithm was actually used. Algorithms must be critiqued in relation to the data they employ, the hardware they run on, and the purpose to which they are tasked. Knowing that a database-retrieval algorithm was trained and run on National Security Agency databases and servers for the purposes of spying on American citizens may result in a very different critique from the one it would receive if it was created or simply used by an artist to make an interactive light installation. For instance, Hadoop, an open-source system for storing and processing large datasets across many cheap computers, has been used by everyone from Netflix to the NSA. Rather than focus on how Hadoop is used by a huge number of companies for various purposes, articles in the *Wall Street Journal* and on Salon.com have tended to focus specifically on how the NSA has used it to create a "government data drag-net."[33] This type of analysis makes Hadoop appear to be primarily useful as a tool of mass surveillance and control. While it is certainly an incredibly powerful and influential piece of code capable of facilitating great harm and worthy of extended critical study, we risk painting with too broad a brush if we do not take into account the various contexts of its use and limit our analysis accordingly.

Algorithms can be used to do everything from simply sorting or adding numbers to deciding which ads you see on Facebook. Even relatively simple algorithms can, intentionally or not, come to manifest particular cultural values in a variety of ways. Like any technology, they can often take on a life and force of their own as they are put to new uses and different users come into contact with them. If a programmer makes a new algorithm that can perform a task particularly well, he or she might choose to sell it or share it with other programmers. To save time (or because they simply do not understand the code), they may simply input this algorithm into their programs without even reading it, let alone exploring its complexity. This is one potential way in which algorithms can take on new meanings and force in different contexts. In addition, Gillespie points out that algorithms are typically tested and trained using databases that can skew what the algorithm generates, and this process can lead to unforeseen, problematic outcomes. For instance, if a face-detection algorithm is trained using a database that features many more Caucasians than minorities, it will likely turn out to be much better at detecting the faces of Caucasians than the faces of minorities. Since many of the free training datasets online that include photos of people are old enough to no longer be covered by copyright and are often primarily filled only with white people, this is a common problem. Intentional or not, this process is a textbook example of how racism comes into being. For instance, skewed testing data led both to Google's facial recognition program identifying African Americans as gorillas and to the Los Angeles Police Department's surveillance cameras far too often identifying them as criminals.[34] Algorithms may appear to be ethically neutral, but they, like any technology, contain the values and perspectives of their creators and users. Indeed, this problem is not new to algorithmic and digital technologies, as photographic technologies have, since their inception, been better at capturing the details of lighter-skinned bodies. As in the case of algorithms, this is in large part due to film companies' testing their products far more often on Caucasians and their concomitant lack of interest in minority customers.[35] These facial recognition systems reflect their creators' lack of care and attention toward minorities, as well as their assumption that such care is not economically advantageous and therefore not worth their time.

In a similar vein, as Safiya Umoja Noble has argued, the industries that use automated recommendations generally display a complete lack of economic and moral interest in minorities, which their algorithms manifest.[36] For instance, instead of improving its facial recognition software to better recognize black people, Google instead chose to just make it so that its algorithms could no longer identify gorillas. While its program now no longer identifies black people as gorillas, it also does not identify gorillas as gorillas.[37] If this problem affected white people, it is safe to assume that Google would have addressed it with a bit more care.

Similarly, in 2016, Amazon began excluding predominantly minority neighborhoods in many American cities from its Prime free same-day delivery service while it extended the service to more white neighborhoods.[38] Numerous articles presented this decision as a continuation of redlining practices that had historically excluded these residents from the creation of public infrastructure and services. Yet Amazon stated that its decision was not racist and was instead based on algorithmically generated recommendations that did not take race into account but, rather, were solely focused on where they could or could not make a profit. While the creators of facial recognition software blamed the results on biased test data, Amazon argued its automated decision was based solely on neutral economic data concerning where the company could make a profit with its service. As many remarked, this response ignores the countless ways economics and class are shaped by racism. This logic serves to make Amazon appear extremely objective for relying on its ultra-rational and ostensibly objective computers rather than human subjectivity to come to conclusions while it also distances Amazon from these decisions. Gillespie argues that corporations often use algorithms as a "talisman to ward off criticism" and "explain away errors and unwanted outcomes."[39] When journalists began reporting on Amazon's decision, the company quickly changed course, but this only illustrated Amazon's apathy toward issues of racial equity—until its decisions raised the ire of its white urban customer base.

The relationship between algorithms and human agency is complex but crucial. Ted Striphas, Wendy Hui Kyong Chun, Alexander R. Galloway, Ed Finn, and many other scholars of algorithmic culture argue that while humans were once privileged with the ability and agency to affect culture, we now share this ability with algorithms. Galloway goes so far as to argue that "the point of power . . . resides in networks, computers, algorithms, information and data" rather than in humans themselves.[40] Yet Matthew Fuller and Andrew Goffey make clear that figuring out where control, agency, and values rest in relation to algorithms is no easy task: as an automatism that invisibly, instantaneously, and continuously works behind the scenes, an algorithm "acquires an agency all of its own."[41] Whether or not algorithms actually have agency, the debate itself illustrates how we too often represent algorithms as being divorced from their creators and the ideologies and values they espouse. Like Striphas, Scott Kushner argues that algorithmic culture sorts, classifies, and hierarchizes "people, places, objects and ideas using computational processes" in ways that oppose and replace elite culture.[42] However, when we look at the role algorithms play in our everyday lives, it is clear that this opposition and replacement of elite culture is incomplete at best.

Indeed, like Jonathan Roberge and Robert Seyfert, I would argue that, in practice, the reality of algorithms is quite messy. They may disrupt the reproduction of elite (white, male, straight) culture or reinforce it—or both at the same time. I try here to study the deep historical context for the production and use of these technologies in order to question how we can "develop an understanding of algorithmic culture that takes meaning seriously by being especially attentive to its inherent performativity and messiness."[43] But while Roberge and Seyfert are predominantly interested in how different algorithms create equally diverse effects, I am also interested in how, regardless of the intentions of the algorithms' creators, individual users necessarily interpret and use them in unique, unintended, and even subversive ways. Whereas Galloway suggests that power is located in algorithms and computers, I argue that power exists in the dialectical relationship between users, algorithmic technologies, and the industries that employ them.

Indeed, humans are still ultimately responsible for what our technologies do, even—or perhaps especially—if we do not understand our creations or are just not paying attention to them. As Imani Perry argues, unintentional (or, as she calls it, "post-intentional") racism and other forms of injustice and discrimination often manifest in inadvertent forms.[44] These problems are not unique to algorithms, and algorithms did not create them. While they may exacerbate the situation, they are merely symptomatic of larger forms of injustice in our culture. As many of the examples throughout this book illustrate, algorithms are not omnipotent; they may malfunction or be misused in various ways, thereby demonstrating the biases that underlie the technology. But for every problem that does come to light, it is impossible to know how many remain hidden.

Recommendation Systems

Recommendation system algorithms hold a special place in the field of digital algorithmic cultural scholarship. Gillespie has referred to recommendation algorithms in particular as a type of "public relevance algorithm" that produces and certifies knowledge: "That we are now turning to algorithms to identify what we need to know is as momentous as having relied on credentialed experts, the scientific method, common sense, or the word of God."[45] In turn, Siva Vaidhyanathan describes the early pre–recommendation system era of the World Wide Web as a space where "clutter and confusion reigned. It was impossible to sift the valuable from the trashy, the reliable from the exploitative, and the true from the false."[46] Similarly, Finn calls recommendation algorithms "magical," especially in the way they can transform our thoughts and actions.[47] Each of these scholars ascribes biblical importance to the emergence of algorithmic culture and the order that algorithms brought to the chaotic anarchy of the early internet. In this view, recommendation systems have supplanted the word of God.

While quite powerful, the chore that recommendation systems accomplish is simple. The systems I will be discussing in this book, which are by far the most commonly used, are made up of content-filtering and collaborative filtering algorithms. These algorithms now make up an important part of modern machine learning and artificial intelligence programs. Dietmar Jannach et al. explains this technology through the example of an online bookstore, where after searching for a particular book, "in one area of the web page possibly called 'Customers Who Bought This Item Also Bought,' a list is shown of additional books that are supposedly of interest to you. If you are a regular user of the same online bookstore, such a personalized list of recommendations will appear automatically as soon as you enter the store."[48] This web page may have generated these recommendations by using content-filtering algorithms, collaborative filtering algorithms, or both. Content-filtering algorithms find other books that are similar in content to the one the user searched for (by the same author, in the same genre, with the same year of release, or with a similar title). This type of system assumes that if users like one mystery novel, they might like another. Collaborative filtering algorithms, by contrast, find other users with the same interests and display what they liked as recommendations. Jannach et al. proposes, "If users shared the same interests in the past—if they viewed or bought the same books, for instance—they will also have similar tastes in the future. So, if, for example, user A and user B have a purchase history that overlaps strongly and user A has recently bought a book that B has not yet seen, the basic rationale is to propose this book also to B."[49] This type of filter does not rely on any information about the object itself and instead only requires information about other users who bought, searched for, or even simply looked at it. Together, these relatively simple filters play a crucial role in structuring much of the internet, contemporary culture, and our lives. They encourage us to think of everything we do in terms of what other people are doing and privilege personal preference and relevance above all else.

Perhaps not surprisingly, these technologies were originally designed to make our businesses, rather than our consumer culture, more productive and efficient. In 1992 the Xerox Palo Alto Research Lab first experimented with collaborative filters as a way to better sort important and interesting emails and other electronic documents from junk.[50] The resulting program, Tapestry, was somewhat like a database that users could manually search through for emails or most any other type of data. The system kept track of every user's actions and searches and used this information to sort the data for relevance. If a user searched for all emails coming from his or her boss, the system would note which of the results were most relevant based on how other users rated or otherwise interacted with the email (i.e., whether they read it, immediately trashed it, labeled it as spam, replied to it, or forwarded it).

Even at this very early moment, users were "inundated by a huge stream of incoming" emails, and Tapestry promised that collaborative filtering could solve this "electronic mail overload."[51] In the *Communications of the ACM* article that introduced Tapestry to a broader engineering industry, the creators used the image of an office worker to illustrate the differences between having no filter, having various simple filters, and having collaborative filtering. With no filter, the worker is buried under piles of documents that rain down from tubes over his head. Using simple filters, the documents still rain down, but from a smaller area. In contrast, the worker with collaborative filtering has a clean desk and his documents are handed to him in a neat stack by another person. Paradoxically and perversely, the creators of Tapestry presented collaborative filtering as a way to combine the strengths of automation with the instincts and care of people by actively "involving humans in the filtering process."[52] Yet "humans" in this context really only meant automatically collected data about human usage of the program, and thus humans were treated only as sets of quantitative data.

While collaborative filters began as a way to make email inboxes less cluttered and work more productive, they have since been used to make all types of digital information and life more orderly. These filters are typically viewed simply as a harmless and pragmatic tool that makes the World Wide Web usable. Yet the ideological, interpellative functions of algorithms do have effects, and every once in a while they become visible. For instance, in 2009, the algorithms that list and rank the sales of books on Amazon "mistakenly" listed all books with gay themes as "adult" materials. Even extremely popular novels like Annie Proulx's *Brokeback Mountain* were suddenly stripped of their sales rank, which resulted in their being removed from product searches, lists of popular titles, and individual user recommendations. This relisting made it more difficult for these books to be found on the website and resulted in *A Parent's Guide to Preventing Homosexuality*'s becoming the top result of a search for the term *homosexuality*. As if obfuscating queerness were not enough, these recommendations actively pathologized it. After discovering this, a large group of users on Twitter used the subject "#amazonfail" to respond angrily with charges that Amazon hides "material it finds distasteful or that it thinks some customers will find distasteful."[53] While Amazon quickly apologized for what it called a "ham-fisted cataloging error" and restored the sales ranks of its LGBT titles, it never explained whether this algorithmic change was caused by a lone homophobic or clumsy programmer, imposed by a hacker, or decided by committee.[54] This is but one well-documented example of how the recommendation algorithms that enable the searching and categorization of the World Wide Web shape and foreground certain types of selves and identities and deny their representation. Everyday uses and effects of recommendation systems, however, largely go unexamined.

While headline stories like #amazonfail are rare, users are constantly presented with recommendations for things they do not want, they already have, that make no sense, or that are outright bizarre. Collaborative filters can lead to seemingly odd and idiosyncratic suggestions. On Amazon, one user created a list titled "Strange Amazon Recommendations" that included, among many other things, a pair of Levi's Men's 550 Relaxed Fit Jeans, which were recommended because the user "owned *Star Wars, Episode III—Revenge of the Sith*," an "OXO Good Grips Salad Spinner," and a Panasonic "Nose and Ear Hair Groomer."[55] While nothing obvious connects these items, Amazon's collaborative filtering algorithms found a correlation between them based on the taste of "similar" users. Amazon's algorithms have also recommended that those who purchase baseball bats also buy a black facemask and brass knuckles and that those who buy the first season of *Babylon 5* also consider *The Complete Guide to Anal Sex for Women*. Perhaps the most famous example of a fortuitous if unexpected recommendation is Target's ability to predict if and when its shoppers will have a baby based on an amalgamation of purchases, including particular types of hand lotions, large bags, and vitamin supplements. One shopper's father discovered that she was pregnant based on the advertisements she was receiving in the mail (he was not pleased).[56]

On the one hand, such wacky anecdotes point to the strangeness of recommendation technologies themselves and thereby give us an opening for analyzing the assumptions and ideologies that they are built on. On the other hand, this wackiness also perversely makes collaborative filtering algorithms appear benign and ideologically neutral. Automated recommendations are so common now that we only notice them when they are bizarre, which can be a sign either of how off the mark they are or, conversely, of what they reveal about ourselves that we do not wish to know. While the gender, race, class, age, location, and politics of user A and user B might be completely different, these technologies unite them around an often invisible personal taste that may appear to transcend diversity and visible differences. Indeed, much of both the justification for collaborative filtering's use and its popularity comes from the inclusivity made possible by its sole focus on consumer history, which means that its calculations are not qualitatively judgmental except insofar as it can only judge those who possess a consumer history.

Yet, even beyond the fact that collaborative filtering reinforces certain taste identities, companies can also skew their algorithms in any of a number of ways to guide users to certain items or brands over others. For many companies like Netflix and Amazon, it is common practice to recommend items created either by them or by others that heavily advertise through their services. Christian Sandvig has referred to this practice as "corrupt personalization" and argues that companies do this to trick us into liking things they have no prior reason to think we would.[57] Thus, recommendation systems do not just reify our identities and preferences but also at times work to transform them, and often it is impossible to tell when this is happening. There are two troubling questions here: How can these recommendations be manipulated, and how do they steer behavior? Rachel Schutt, a senior statistician at Google Research, has argued, "Models do not just predict, but they can make things happen." Steve Lohr, in the *New York Times* article in which Schutt is quoted, elaborates, "A person feeds in data, which is collected by an algorithm that then presents the user with choices, thus steering behavior."[58] While, rhetorically, the ability to choose becomes largely synonymous with freedom, digital technologies guide these choices through recommendations to shape user behavior.

The Limits of Algorithmic Power

Much of the popular and academic discussion about recommendation systems tends to assume that these systems can (or at least shortly will) be able to guess what we like with pinpoint accuracy and thereby also know us intimately. Popularized by films such as *Blackhat* (2015) and *Eagle Eye* (2008), television series including *Person of Interest* (2011–2016) and *House of Cards* (2013–2018), and novels such as Dave Egger's *Circle* (2013) and William Gibson's *Pattern Recognition* (2002), these discourses encourage the paranoia that these technologies may soon learn so much about us that they can accurately predict not just what we desire but also what we will do. Even more, we fear that these capitalist corporations will not stop at simply knowing our futures but will also try to shape them to their will.

This fear seems to be justified, given that these industries appear to highly value conformity and control. Yet I argue that while automated recommendations can be very dangerous and have dire consequences, we can and do continually read them against the grain, interpret them with a campy sensibility, subvert them, ridicule them, or simply reject them outright. Algorithms are neither transcendent nor simplistic; they are the messy product of many drafts and revisions by countless engineers working for companies with continually transforming objectives, products, and consumer expectations. If their algorithms are developed through machine learning techniques, engineers may have no idea how exactly they work. Even when dealing with simpler technologies, as in any work environment, misunderstandings and a lack of communication are common and lead to algorithms that do not always do what was intended.

Take, for example, Neiman Marcus's current recommendation system, MyNM, which launched in 2014. Fifty-three years after the IBM Gift Advisory System, MyNM continues to illustrate many of the same anxieties and cultural debates about exclusivity, class, surveillance, gender norms, and the automation of jobs and relationships. Like the Gift Advisory System, MyNM uses personal information about users' shopping habits and interests to generate a list of items that those users might like to purchase. Yet while the Gift Advisory System assumed that shoppers were buying gifts for others, MyNM makes recommendations for the shoppers themselves. The site, like all contemporary recommendation systems, presents itself as a tool for learning more about your own tastes rather than those of others. Moreover, MyNM seeks to bring users' attention to new items or items that they may not have seen before, seemingly regardless of whether it receives any data indicating the users would like the new products. MyNM prominently presents users with images of new products that have arrived since the last time they shopped online, items on sale, and items that have been mentioned the most on sites like Pinterest and

Twitter. "Liking" or purchasing items on the site will theoretically transform the list, allowing it to continually show you new items that more and more often reflect your interests and identity. Clicking on these images opens a "flipbook" that approximates the original Neiman Marcus catalog, but with $600 sweatpants and Ferragamo flip-flops rather than "his and hers" sarcophagi. Instead of helping the customer find the "perfect gift," MyNM reflects back to users a "better" version of themselves "enhanced" by Neiman Marcus products. (In later chapters, I will discuss how many industries, including self-help and online dating specifically, present their recommendations as ways to improve yourself.)

While in 1961 (and even 2010) this automation of recommendations was a novelty, by 2014 it had become common—if not necessary—for all major American retailers. Since Amazon and eBay helped popularize the technology in the early 2000s, recommendations have become an obligatory and central component of e-commerce, which has rapidly become an important part of the American economy. This transformation affects mainstream and luxury goods retailers alike. In 2015, Amazon—which at that point existed only online—became America's most valuable retailer. *Fortune* magazine and others have attributed much of Amazon's growth to its use of recommendation systems and the personalization it affords.[59] Such articles assert that these systems make shoppers feel more comfortable by making the act of searching through a website seem more like browsing in a store. In 2014, after MyNM launched, Neiman Marcus made 21 percent of its sales online. Ginger Reeder, the vice president of corporate communications at Neiman Marcus, explained the personalization of MyNM as "an attempt to replicate the personal relationships that are the hallmark of our customer/associate relationships in our stores."[60] Neiman Marcus has long advertised its personalized service as a way to justify its high prices—a quality that cannot be easily translated to the largely impersonal environment of e-commerce.

Indeed, in contrast to the face-to-face relationships created between shoppers and sales associates, MyNM recommendations have their limitations. For instance, after I added a number of suits, polo shirts, and ties to a list of my favorite items, the site continued to primarily recommend dresses, high heels, and handbags to me; it clearly does not take gendered preferences into account and assumes a female audience. And given that the models on the site are exclusively white, it also does not appear to take racial variation into account except through its exclusion. As with the facial-recognition algorithms, it is possible that these recommendation algorithms were only tested on white bourgeois women, Neiman Marcus's core demographic. Even as it markets the personalized aspects of this site and the relationships it can create between the shopper and the store, the site itself appears to present the same items to everyone, regardless of their preferences or personalities.

In many ways (and I cannot stress this point enough!) the marketing around these recommendations is more important than the recommendations themselves. Online, even when they do not work well—and often they do not—recommendations become symbols of luxury, care, and personal relationships. MyNM is just one of many recent sites through which high-end retailers have begun to associate recommendations with luxury. Bloomingdales, Saks Fifth Avenue, and Barneys are all "engaged in an arms race to keep offering finicky luxury shoppers the next big thing in e-commerce."[61] This "arms race" frames the ability to receive recommendations as a sign of privilege and class while it also frames these shoppers as "finicky," or overly exacting and demanding concerning what they want. The Luxury Daily blog, "the news leader in luxury marketing," argues that "a personalized experience is an effective way to attract and retain affluent consumers who seek an individualized shopping experience."[62] This statement implies that personalization technologies and the recommendations they generate convey a sense of individual attention to each consumer even as they address all consumers in more or less the same way. These recommendations, whether they work or not, are created through vigilant surveillance of customer activities, which is then paradoxically represented as a form of care. Even when the recommendation system "fails," its very presence demonstrates that Neiman Marcus cares about "you."

While it is tempting to argue that MyNM is just an example of a bad recommendation system and that over time it will invariably become more accurate, I argue throughout this book that we should understand the inaccuracies of these systems not simply as faults that must be fixed but rather as a systemic and essential part of the experience of recommendation systems; they may be errors, but they are also affordances that reveal much concerning the role of recommendations in our contemporary culture.

This book is organized around different areas of contemporary culture, including social networking, digital media, online dating, and cosmetic surgery. I examine debates that surfaced concerning the role recommendations play in how we represent ourselves online through social networks; how our online experience is shaped by our media-viewing and news-reading habits; how our interpersonal relationships are shaped and algorithmically constrained; and how technologies condition us to imagine and transform our bodies in accordance with current ideals of beauty. Together, these examples give a fuller account of how recommendations and algorithms have come to play a central role in every day and every moment of our lives.

I start in chapter 1 with a cultural history of the "recommendation" and the debates about agency and temptation that have long surrounded this rhetorical figure. I then explore how this debate shaped the imagined uses and fears of early computers and many of the more troublesome aspects of contemporary culture. Indeed, I argue that the recommendation has become a central technique of neoliberal, postfeminist, and postracial discourses and the culture they generate. As I illustrate through a discussion of the GroupLens Research Project at the University of Minnesota, one of the first and most influential early experiments with collaborative filtering, these tensions between personal agency and social control in contemporary digital culture provided the impetus for recommendation systems.

In chapter 2, I discuss the birth of social networking and digital recommendation technologies in the mid-1990s and their relationship to the lives of female professionals. Most popular histories of these technologies portray them as being primarily created by and for men, but many of them were actually pioneered by Pattie Maes, a professor at the Massachusetts Institute of Technology (MIT). In the MIT Media Lab, Maes created several digital recommendation systems and implemented them in various programs, including one that could automatically schedule meetings for users, a program for prioritizing and sorting emails, one of the first matchmaking programs, and Firefly, one of the first online social networks. These technologies and their intended uses have played a central role in the growth and development of the World Wide Web Consortium (the organization that governs and maintains the web) and are responsible for much of the popularity and economic feasibility of sites like Amazon, Facebook, Google, and Netflix.

As I demonstrate, Maes developed these technologies specifically to help female professionals like her manage the many difficult everyday choices that having a family and a career often entails. MIT during this period was an intensely male-driven institution that many felt was not welcoming to women. While Maes started out as a researcher of artificial intelligence and robotics, her experience at MIT and of managing her life and new family led her to use her skills to develop technologies that she hoped would alleviate her many everyday dilemmas. In the process, she used recommendation systems to do everything from managing her schedule to keeping in touch with family—dilemmas that men experience but that continue to be heavily gendered in problematic ways. Her work on recommendation systems also led to innovations in user privacy protections, and she became an important voice in the struggles over agency, surveillance, identity, and privacy that continue to pervade digital rhetoric. This history and a discussion of Maes's influence show how central gender, race, and class politics have been to the growth of digital technologies.

In chapter 3, I discuss how current scholarship and our culture more generally have vastly overemphasized the power that algorithms have over us by exploring cases in which users have explicitly reacted to the recommendations that algorithms generated for them. These examples come from interactions with TiVo, Netflix, and Digg, three media distribution companies that helped popularize recommendation systems. In so doing, their actions have raised many digital privacy and surveillance concerns, particularly in regard to gender and sexuality. I address how media distribution has long played an important role not just in constructing the public sphere but also in defining how citizenship is imagined and practiced. This influence—and its relationship to postfeminist, postracial, and neoliberal ideologies—has recently been most evident in controversies over how recommendation systems have been used to define what user information is public and what is private. Indeed, there has been a great deal of pushback against these efforts, and I consider how users do not simply adopt the online subjectivities offered to them but often subvert these offers. I start by looking at TiVo in relation to allegations concerning its role in sexual harassment and the "outing" of homosexual audiences. While this event brought to light the intense forms of surveillance that recommendations necessitate, I focus on the tactics of users who mislead this surveillance for their own purposes. I then discuss Netflix and the ways in which its recommendations have been shaped by concerns over sexual privacy protected by the U.S. federal Video Tape Privacy Protection Act (1988), a sweeping and influential law that developed out of the Robert Bork Supreme Court hearings. This example illustrates the various reading strategies we bring to algorithms— whether dominant, negotiated, or oppositional—and how our interpretation and use of algorithmic data are always historically situated. I then consider Digg, an early social news site that played a pivotal role in defining how digital citizenship is practiced by celebrating the commercialization and individualization of the public sphere. I specifically discuss how the users of Digg rebelled against the site when its underlying algorithms changed in ways they did not like. This fight over Digg's algorithms ultimately led to the downfall of the site itself. In each case, the question of user rights on these sites and the status of digital communities within a contemporary culture intersect with the rapidly changing relationship between identity and privacy.

In chapter 4 I consider how algorithmic recommendations work (or, more often, do not work) on online dating sites. I begin by exploring how computerized matchmaking was used early on in the 1950s and 1960s to create couples as a response to changes in dating patterns that resulted from more women working and moving to urban areas. I then discuss the matchmaking questions and tests on eHarmony in relation to Chemistry.com and OkCupid, three of the most influential sites known for their recommendation, or "matchmaking," systems. These three sites algorithmically model users and attraction in disparate ways that together illustrate that—while much of the discussion about dating websites focuses on which one is the best or most accurate—algorithms are shaped by particular assumptions about what constitutes one's "identity" and how to define it. While eHarmony's test focused on the personality and psychology of users, Chemistry.com tries to divine your genetic and chemical makeup in order to find your perfect biological match. In contrast to this phrenological tactic, OkCupid's matchmaking system relies on random questions proposed by users that were statistically shown to lead to successful relationships. None of these sites has been shown to be particularly effective in helping people find their soul mates. However, while they may not actually help you find love, their algorithms do enforce traditional gender norms—even as they also make dating a safer and more empowering experience for women. These two impulses have influenced the ways in which dating sites function. Moreover, these sites' use of questionnaires and recommendation systems to manage the sexual, racial, class, and gender expectations of users reifies particular conceptions of a "successful relationship." Even so, I also address how users parody, critique, play with, and otherwise challenge these sites and their underlying assumptions concerning definitions of love and identity. I end by discussing various parody dating sites, including Cupidtino (a site for Mac lovers), H-Date (a site for those with herpes), Old People Dating, and OKMudblood (a Harry Potter–themed site "for wizards and various crossbreeds"). These sites parody the insularity of dating websites and the narcissistic assumption that we can only find happiness in our mirror images.

In chapter 5, I broaden my discussion of how recommendation systems and contemporary culture frame the body and embodied attraction in digital media further by discussing cosmetic surgery websites such as Anaface, ModiFace, and Make Me Heal. While many of the sites I discuss in this book argue that their recommendation systems mirror their users' desires and sense of self back to them, those used in the context of cosmetic surgery demonstrate that recommendations more generally ask us rather to "misrecognize" ourselves in their recommendations. Many of these sites emphasize ways in which recommendation technologies can help users "rate" bodies and often automatically propose ways to change one's own shape and appearance. I start by examining how cosmetic surgeons in the 1980s used photography to standardize both beauty ideals and their larger industry. Now, cosmetic surgeons use digital photography to help patients imagine their post-op bodies, but still in highly constricted and standardized ways. I examine how contemporary paradigms of choice and individuality transform the human body through websites and technologies that analyze, judge, and rate a person's appearance in order to recommend clothing, makeup, and plastic surgery operations. Throughout, I show how these recommendation technologies and their varied uses transform and complicate notions of what our relationship to self-representations consists of in our current media landscape—and what this relationship could become. I look at a large number of sites and technologies that all fall back on a long history of regulating bodies through the use of media and recommendations that privilege the white, American body as a timeless form of beauty that all must live up to. At the same time, these systems often allow users to experiment with their physical form and see what a "bad" surgery could turn them into. While these sites encourage this play, often to convey the cautionary tale of what would happen if you went to a different surgeon, I look at how many people nevertheless use these tools to reassert their control over their bodies and aesthetic judgment. These automated recommendations and transformations of the body help to construct the body as a product of choice and make it an important site of an empowered consumerism that is rapidly enforcing the commodification of attraction.

I conclude by discussing ongoing efforts to both make more critical, ethical, and socially conscious recommendation systems and find resistant uses for them outside the consumer sphere. I ask whether recommendation systems must be a tool of control, or whether they can also be put to emancipatory purposes, and I come to no easy answers. I argue throughout this book that the digital recommendation system has become a primary technology for generating and maintaining the postfeminist, postracial, and neoliberal strains in our culture, which now define choices as burdens better left to free enterprise and automated technologies to enact. Digital recommendation systems play a role in every area of life and help shape every choice we make in a digital environment. The contours of these choices follow a logic that pervades all areas of life and does not recognize a distinction between the deeply personal and the overtly communal, nor between freedom and oppression. Like postfeminism, postracialism, and neoliberalism more generally, these technologies purport to be blind to race, gender, class, age, and ability. Yet I believe that this study makes it clear how thoroughly these embodied marks of culture continue to pervade the digital sphere through the ways in which technologies frame our choices and, as a result, come to shape and limit the very contours of what we deem possible—and how we continue to reach beyond them.

1

A Brief History of
Good Choices

•••••••••••••••••••••

In *Existentialism Is a Humanism*, Jean-Paul Sartre tells of how one of his students came to him asking for advice. It was 1940, and the student's brother had just died in World War II. He yearned to avenge him by joining the fight, but he also had a grief-stricken mother who needed care. In response to this dilemma, Sartre responded, "You are free, so choose; in other words, invent. No general code of ethics can tell you what you ought to do: no signs are vouchsafed in this world."[1] While this comment may seem ineffectual and perhaps even flippant, Sartre was a strong believer in the importance of human agency, and any student asking him for advice should have expected just such a response. Reflecting on this episode, Sartre noted that when we ask people for advice, we tend to already know their perspectives and what advice they will probably give (i.e., if the student had actually wanted a more direct answer, he should have asked someone else). As Stephen Priest argues, "To choose an adviser is to make a choice. It is also to choose the kind of advice one would like to hear."[2] No matter how much we try to divorce ourselves from the many small and gargantuan choices we must continually make by relying on recommendations from those we trust, the decision is always ultimately ours. According to Sartre, this freedom to make decisions and use such choices to continually invent ourselves throughout our lives is less a gift than a burden that we cannot escape; "man is condemned to be free."[3]

The question of how constrained one's actions are by social structures and how, when, and whether individual agency can really guide our behaviors lies at the heart of much of the humanities and social sciences. While René Descartes's famous phrase, "I think, therefore I am," idealizes human agency above social structures, others, like Karl Marx, argue that our choices are largely guided by the ideological structures around us. By arguing that we are free and must invent ourselves accordingly, Sartre seems to sit at the far end of this debate. Yet his most noteworthy move here transforms our ability to choose from a pleasurable strength to our greatest liability and the source of our misery and anxiety. Choice and freedom are typically presented as universally desirable and thereby also as the neutral premise of all post-Enlightenment ideologies and the formation of a rational cohesive subject. However, for Sartre, the privileging of freedom and choice is itself the ideology that must be critiqued and deconstructed. How rational can people really be if the freedom they seek is a source of their consternation?

While I am inspired by Sartre's move to consider choice a burden, I would argue that he actually does not go quite far enough in his critique. For Sartre, while no choice is unassailable, the very act of making choices for oneself still remains a central expression of agency and authenticity. Yet the question of what constitutes a choice is far more subjective than he indicates. What one might call a choice, another might experience as a compulsion, obligation, or ultimatum. As contemporary activism on both the left and right, from Black Lives Matter and Me Too to efforts to "build the wall" and "make America great again," illustrates, poverty, racism, sexism, and other forms of exploitation and exclusion limit the number, type, and quality of choices many people can make. For some, making certain choices subjects them to greater scrutiny; circumstances may even make it difficult to perceive some people as being capable of making choices at all. By arguing that we are responsible for all our actions and their outcomes, Sartre ignores the structures within which choices must be made by those who are more constrained in the options available to them.

Indeed, much of contemporary racism and sexism does not just limit the options certain people have but can also turn those things they do experience as choices into forms of constraint. For example, Linda Duits and Liesbet van Zoonen argue that neither the decision of Muslim young women to wear headscarves nor the decision of white young women to wear G-strings is typically represented as a choice, even though each is often experienced as such by these women.[4] They argue that these are just two examples of how the choices of young women are rarely respected as choices and that young women are therefore rarely thought of as having agency at all. Rather, they are presented entirely as subjects only capable of expressing the will of external influences. Celebrating choices continues to privilege bourgeois white men, who typically have access to more and better options from which to choose. Indeed, choices do not lead to agency and freedom; rather, individuals must first be imagined to have agency and freedom in order to have their decisions count as choices.

The relationship between choices, freedom, and influence is complex and culturally situated. Here, I am interested in exploring how the recommendation—which simultaneously, if paradoxically, privileges individual agency as well as structures of control and influence (which typically appear to limit agency)—can help us critique how choice currently functions. I begin by comparing the etymology of *recommendation* and *suggestion* in order to elucidate current debates about whether outside influences and advisers lead to a perversion or affirmation of personal choices. I then focus on how this debate about conceptions of choices and recommendations influenced the concurrent and overlapping development of both artificial intelligence (AI) and feminism. Historically, much of the discourse concerning the internet and digital technologies has focused on how they enable users by broadening their horizon of possibilities and enlarging their sense of free will. Yet this perspective ignores the multitude of ways in which digital technologies present choice not as liberatory but rather as a burden. I then address how this history has shaped currently intertwined neoliberal, postfeminist, and postracial ideologies, and how these ideologies now frame choice as a burden. Indeed, since the mid-1990s, various industries have used recommendation systems to profit from presenting choices and agency as burdens in order to then sell us a sense of relief in the form of convenience and simplicity. Digital media continually prescribe predefined choices and actions through their algorithmic interfaces of drop-down menus, buttons, check boxes, and other clickable objects. Unless you have a great deal of coding experience and time on your hands, these are typically your only options. Of course, you could always just reject all the choices by not using particular seemingly omnipresent websites and applications (or, for that matter, your computer), but due to economic and personal responsibilities and expectations, that option rarely feels plausible. Regardless, these limited choices are often actually presented positively as a feature of digital technologies rather than their greatest limitation (or, perversely, limitations are now the affordance). For instance, Apple markets its design principles on simplicity, a code word for limiting the number of options its customers are presented with in order to avoid confusion and making those few options they do have appear as desirable as possible. Tracing the long history of recommendations helps to explain how these contradictory discourses and impulses work today.

From even the earliest moments of the World Wide Web, recommendation systems were used to sort through the abundance of information and choices that digital culture created. In my introduction I discussed how Tapestry, the first collaborative filtering-based recommendation system, was presented as the antidote for too much information, too much clutter, and too much choice. To explain how recommendations became a major industry, I end this chapter by recounting how the GroupLens Computer Science and Engineering Research Lab at the University of Minnesota, one of the earliest groups to work on recommendation systems in the early 1990s, conceived of and eventually marketed these technologies. Their work continues to resonate through a vast industry that sells recommender and personalization technologies to companies world-wide. At the same time, I argue that by framing free choice as a burden best alleviated by capitalist forces, GroupLens and the many companies that followed them also frame the act of making unrecommended choices not just as dubious but also as subversive—with both the negative and the positive connotations that *subversion* may entail. This alternative understanding of choice has been deeply inflected by the role that choice and recommendations have played throughout the early history of computing. While Sartre spoke of choices making us who we are, now these industries frame unrecommended choices simply as a waste of time and resources—that is, the ultimate sin, according to capitalists.

Recommendations and Suggestions

While I use the term *recommendation* throughout this book because that is the most common term used by software engineers and digital industries, some also use *suggestions*, *top picks*, and other terms interchangeably. These terms are all now synonymous, but each has distinct connotations and a long history that illustrates current values, debates, and anxieties about algorithmic recommendations. As Raymond Williams argues, the meaning of words may change, but the history of these transformations often haunts them into the present: "Earlier and later senses coexist, or become actual alternatives in which problems of contemporary belief and affiliation are contested."[5] In our contemporary use of *recommendation*, *suggestion*, and *top pick*, we can see the traces of larger battles over the shape and meaning of digital culture.

For instance, Netflix (which I discuss in detail in chapter 3) had long labeled the movies and series it highlighted as "recommendations" but has recently begun calling them "top picks." At the same time, it also began advertising that it was no longer just using algorithms to make recommendations but had also hired many people to watch all its media and label them with keywords to improve its picks.[6] Here, "top picks" implies that someone, instead of some machine, is physically making a pick for you. This distinction foregrounds contemporary debates about the value of the labor of AI versus that of humans, raising the question of whether we care if the advice we get is in some small way shaped by a human, and whether we trust humans or machines more with all the personal data that go into making such "picks."

Pundits and academics alike continually debate how digital technology and culture figures "new" concerns about choice, privacy, security, agency, surveillance, and relationships, but here I will examine how these anxieties have always been implicit in the way we use two very old and very connected terms: *suggestion* and *recommendation*. Our anxieties over taking advice and ceding agency are a result not of digital technologies but rather of the way we rely on them for such advice. *Suggestion* and *recommendation* have distinct histories that illustrate how our fears, hopes, and misgivings about these actions have played out over the centuries and into the present. Indeed, such fears have always been a central aspect of suggestions and our ability to make up our own minds.

Suggestion first came into English in the mid-fourteenth century, and its original meaning was "prompting or incitement to evil; an instance of this, a temptation of the evil one."[7] Rather than describing a benevolent act, this definition points to the lack of self-control and agency of the one receiving the suggestion, as well as the evil intentions of the one who suggests, who, in this context, was understood to be the devil. Most definitions for *suggestion* focus on questions of temptation, self-control, and harm. While Sartre places the burden of choice squarely on the subject, suggestions historically have caused confusion about where agency lies. At times, suggestions stress how rarely we actually have final say over any choice we make. In John Milton's *Samson Agonistes*, suggestions are depressed—if not suicidal—thoughts that "proceed from anguish of the mind and humors black."[8] Here, the uncontrollable unconscious and bodily chemistry become the sources of suggestions and the choices that spring from them. We may make such choices, but only when we are not ourselves. In the works of Geoffrey Chaucer and Thomas Kyd, suggestions tend to be false, and subjects must be "freed from" them through confession.[9] Not only evil, these suggestions supplant our sense of personal choice and agency. Further, these examples illustrate that it is within our power to reject even these suggestions, but only if we confess and side with God.

By the beginning of the twentieth century, *suggestion* had become synonymous with hypnotism and the ability to take full control over another's thoughts, actions, and desires.[10] While we now use this term in a much more neutral fashion, it still connotes anxieties about self-control and manipulation. We fear that the suggestions that now surround us online will make us suggestible, or easily led toward bad or even evil decisions. For example, when Facebook announced that it was experimenting on its users by sometimes making depressing or uplifting stories appear more often on their feeds (which is Facebook's way of suggesting articles that you might like), many, like Jim Sheridan, a member of the British House of Commons, feared that such actions would "manipulate people's thoughts in politics or other areas."[11] Clay Johnson, the cofounder of Blue State Digital, the firm that built and managed Barack Obama's presidential online campaign platform, questioned whether such suggestions could control people en masse to start revolutions and swing elections.[12]

If Facebook started only showing or suggesting positive articles about an elected official or corporation, we may mistakenly interpret this to mean that there are no negative articles; choice is meaningless if we are ill informed and unable to imagine the many choices that we can actually make. With the Trump campaign's candid use of social media to spread outlandish rumors, the transformation of "fake news" into a "national strategy" in the form of Cambridge Analytica, and Breitbart and the ascendancy of the white supremacist Steve Bannon, anxieties over the relationship between suggestions and suggestibility have only become more pronounced. As Ann Coulter puts it in a Breitbart article, "Most people are highly suggestible. That's why companies spend billions of dollars on advertising."[13] Advertising is thus a form of suggestion; *recommendation*, however, connotes something more benign.

If suggestions come from the devil, recommendations often come from God. In the King James Bible, the grace of God "recommends" that Paul and Barnabas sail to Antioch, where they start a church and where the followers of Christ are first called Christians.[14] Recommendations are predictive and connote a placement of faith in the recommender and authority figures more generally. While we have the ability to discard a recommendation, it would be hard to turn down God, and we would do so at our peril. Along with making decrees, writing rules and laws, and doling out punishments, clerics, parents, teachers, the state, and other authority figures have also employed recommendations as a form of ideological control and regulation. While laws and rules become prohibitive sticks, recommendations can function as carrots that place subjects in the position of having to affirmatively do what you want them to. In this sense (and invoking Louis Althusser once more), the recommendation clearly functions as a primary technique of ideological apparatuses and the culture industry.

If suggestions tend to be inventive associations or advice that should not be taken, recommendations historically connote the cementing of associations and advice that should be thoroughly committed to. When *recommend* first entered the English language at the beginning of the fifteenth century, it had two intertwined meanings: "to praise, extol, or commend" a person or thing and "to commit (oneself or another) to a person or thing, or to someone's care, prayers, etc."[15] By praising the person or thing, you commit, or deeply connect yourself, to it in the hopes of also strengthening your relationship with the recommendee. In Jane Austen's *Pride and Prejudice*, Miss Bingley cautions Elizabeth against George Wickham by stating, "Let me recommend you, however, as a friend, not to give implicit confidence to all his assertions."[16] Miss Bingley's recommendation signals a sense of benign authority and elitism that is central to its role as a form of cultural constraint. She offers her recommendation not simply to impart wisdom but also in an attempt to become more of an authority figure to Elizabeth.

The lines between recommendations, suggestions, and other forms of advice are necessarily blurry, and all set up a power dynamic between those who give advice and those take it. Whether in the form of a *New York Times* film review, a "Dear Abby" letter, a *Cosmopolitan* list of ten ways to get tighter abs, or a set of recommendations for how to eliminate racial bias in police departments (or sexual harassment at Uber), such advice purports to come from someone who knows more than the reader; the recommendation often makes the recommender an authority figure, not the other way around. At the same time, there are important distinctions between these various forms of recommendations. One can use a film review as a recommendation concerning whether to see a film, or one can read it specifically so as not to have to see the film at all. A "Dear Abby" letter or *Cosmopolitan* article can provide guidance, or it can be a source of entertainment. A list of recommendations for a police department might appear to be optional but are often understood to be orders.

Digital media further complicate the question of how much force recommendations have and how manipulative they are. For instance, the list of "sponsored" ads on Facebook's right-hand column is often generated by recommendation system algorithms, but they are clearly paid content rather than recommendations from Facebook itself. Likewise, searching or browsing can feel like an empowering experience. However, when one browses what are sometimes labeled as "curated shopping sites" for those with expensive tastes, such as MyHabit (a defunct designer-brand site previously owned by Amazon), ShopStyle (owned by the women's lifestyle brand PopSugar), or Gwyneth Paltrow's lifestyle brand and website, Goop, the horizons of possibility are hidden but impenetrable. Such sites all cater to very specific bourgeois audiences and only carry items that demographic supposedly would purchase or be drawn to, such as portable yurts or 24-karat gold dildos.[17] Such "recommendations" steer shoppers in a particular direction in ways not always apparent.

Indeed, while recommendations have historically connoted the trustworthiness of a benign authority, they nonetheless imply a clearly "wrong" choice. Although people are always free to engage with or ignore recommendations, they are also often made to feel as though they are doing themselves a disservice if they do not take the recommenders—who are positioned as experts—up on what they present as the best choices. Recommendations thus seem to preserve the ideals of free will and individualized choice even as they employ these ideals as mechanisms of social control and constraint. They inspire people to action not by a fear of punishment but rather by the internalized anxiety that if they do not take the advice and things go badly, they will have no one to blame but themselves.

In contrast to most traditional recommendations, such as literary reviews or curated "best of" lists, digital automated recommendations often forgo their air of authority—along with the elitist and canonical issues such authority presents—in favor of greater personalization and familiarity. Rather than present their recommendations as curated by experts, many sites like Netflix and Amazon present their recommendations without any sign of how generally well liked or esteemed the items are. Instead, they just announce that the site thinks "you" specifically would like them based on your previous activities on their sites and, in some cases, on the entire World Wide Web. They imply that the essentialist logic that governs traditional demographic and cultural groupings arranged on the basis of gender, race, age, nationality, and class is no longer the dominant logic for imagining audiences or communities of spectators. Instead, it is all about "you" as an individual. Indeed, rather than aiming at an audience at all, many digital recommendation systems instead focus on specific users, based on a large amalgamation of both personal data related to those users and information on the recommended items.

Yet companies that use recommendation systems only keep track of certain parts of a person's identity while they ignore others, and the parameters of these data are amorphous and proprietary. While the data they use may contain information about a user's class, gender, age, ethnicity, and race, it is often unclear how sites like Google, Facebook, or Netflix use this information to generate recommendations or even whether such cultural markers play any role in their calculations at all. Netflix has stated that it relies on what users view and enjoy in order to make recommendations, but it has neither confirmed nor denied whether it also takes the age, gender, sexuality, or race of users into account. By making this knowledge proprietary, Netflix and other sites like it obscure how our actions affect what we can see and experience online. They also make it impossible to know what other things we could experience if we acted differently.

Scholars have been arguing about whether this is a feature or a bug of digital technologies since close to the beginning of the World Wide Web. In 1995 Nicholas Negroponte described Fishwrap, an MIT Media Lab effort to create a digital newspaper personalized to feature stories that directly appeal to each individual reader's taste. In *Being Digital*, he referred to this newspaper as *The Daily Me* and praised the possibility that one might pay more for ten pages of content he or she definitely wants than for one hundred pages of content the individual only might want.[18] Later, both Cass R. Sunstein and Joseph Turow separately described how these algorithmic technologies have the deleterious effect of placing "us into personalized 'reputation silos' that surround us with worldviews and rewards based on labels marketers have created reflecting our value to them."[19] Such siloing may split our communities along lines of political views, culture, race, gender, sexuality, and class and make it more difficult for us to come into contact with people different from ourselves.

Indeed, the user's lack of ability to clearly tell how sites like Netflix and Amazon use these cultural markers and personal data speaks volumes about how identity is now represented online precisely through its explicit absence. By not referencing specific groups or demographics, these sites appeal to users at the level of the individual, while also suggesting that their service is for everyone. I discuss the political ramifications of this dynamic in more depth when focusing on Digg in chapter 3.

This desire to appeal simultaneously to everyone in the world and "you in particular" at times results in contradictory recommendations. As I discuss further shortly, these moments of "malfunction" generate potentials for surprising effects and unpredictable relationships between particular users and their recommendations. Yet, more often, the capitalist desires of these companies omit the specific circumstances of particular types of consumers who may not be addressed by these algorithms, algorithms that tend to privilege dominant identities: white, male, straight, affluent. Some of the most extreme and flagrant examples of these effects occur in the areas of finance and the courts. Credit-scoring companies like FICO have already started using information on social networks in their calculations. While it is illegal in the United States for companies to discriminate against customers based on race, religion, or gender, companies can consider where customers live and the income and repayment history of their friends and neighbors in making credit score decisions. As Kaveh Waddell argues in the *Atlantic*, this information is heavily correlated with race and can "allow creditors to end-run those requirements."[20] At the same time, courts around the country and globe now rely on algorithms that take in similar data in order to determine the recidivism risk of defendants and criminals for the purpose of setting bail and considering parole.[21] While these risk-assessment algorithms were implemented in order to avoid the possibility of racist judges' imposing stricter rulings on people of color, ProPublica found that these algorithms were themselves quite racist. Rather than leading to a more equitable and equal world—the digital utopia we have been continually promised—these algorithms are instead reflecting and amplifying only the worst aspects of our culture. While these practices are clearly egregious, the same logic undergirds even the most quotidian of recommendation systems.

Although contemporary culture characterizes personal choice as a primary way in which people perform both their freedom and their individuality, the extraordinary popularity and omnipresence of recommendation systems in online environments demonstrate ways in which this freedom is often heavily restrained and controlled in order to make these choices and decisions "easier." At the same time, by recommending only those choices that many others across the world have already made, these systems commodify choice and make us generic. The act of making choices ceases to be a performance of individuality and instead becomes an operation of conformity. Indeed, as I argue throughout this book, recommendation systems and the digital technologies and websites that use them construct personal choice in all areas of life not as a sign of our unique value in the world but rather as a burden best relieved by outside forces and automated technologies.

Suggestion and *recommendation* may currently be synonymous, and associations with God and the devil have become increasingly obscure. Yet both terms continue to encapsulate the many potentials and anxieties related to choice, agency, privacy, trust, and connectivity that they always have. As Amazon continues to recommend Nazi paraphernalia and YouTube's algorithms show children videos of their favorite characters being maimed and murdered, it is clear that algorithms have no inherent morality.[22] Automated recommendations have become a tool for mitigating and managing both monumental and everyday decisions, but they do not necessarily help us make good decisions.

Nevertheless, the "recommended" item generally seems like a good idea. Given its associations with goodness, it is not surprising that the term *recommendation* has become paramount online while the word *suggestion* has all but disappeared from digital nomenclature. Most every site that employs these algorithms refers to their output as "recommendations," "top picks," or a variation such as, "You may also like . . ." The only current example I could find of a recommendation system that prominently uses the word *suggestion* is DashVapes, a company that "suggests" vape flavors based on users' stated preferences; such "suggestions" only make the advertising techniques (and health benefits) of vaping appear more spurious.[23]

Conflating Recommendation and Choice

As recommendations moderate between the idealization of personal choice and the limitations of ideological control, they also serve as a mechanism for the shaping of culture and technology. While recommendations were being automated even during the earliest moments of modern computing, their history intertwined with conceptions of gender and AI. At the same moment that Sartre described the burden of choice, his lover Simone de Beauvoir articulated the performative aspects of femininity in *The Second Sex* (1949) and Alan Turing defined AI (1950). Since then, these three discourses have been intertwined and feeding off each other. Beauvoir articulates gender as a burdensome set of (forced) choices that one must make again and again. Turing genders AI by arguing that it must prove that it is capable of making intelligent choices specifically by convincingly copying gender norms. Together, these discourses treat gender, recommendations, and AI as related algorithmic processes, each dependent on weighing an individualized sense of choice against social norms and control.

The relationship between gender and choice is in many ways parallel to the construction of choice more broadly within contemporary digital culture.

Judith Butler argues that Beauvoir's famous statement, "One is not born, but rather becomes, a woman," is the foundation for the "reinterpretation of the existentialist doctrine of choice whereby 'choosing' a gender is understood as the embodiment of possibilities within a network of deeply entrenched cultural norms."[24] The choice of how to perform your gender or sexuality clearly has a very different cultural status from most any other choice and is subject to far more potential restrictions and reprisals. Even so, I would argue that it still follows the same logic of less weighty dilemmas, such as which book to buy next on Amazon, in that we make both choices based on a combination of considerations, including what we want, what we need, and how we think others will react. And indeed, like all choices, the status of gender performance as a choice is heavily context dependent, and many experience it instead as a compulsion, a mandate, or an order.

For Beauvoir and Butler, one does not make a choice from outside culture, and choice is neither a source of unlimited freedom nor entirely deterministic. Rather, it is a performative and embodied positioning of oneself "within a field of cultural possibilities"; for Butler, choice is always placed in relation to what is recommended.[25] Butler argues that while people always and continually make choices "to assume a certain kind of body, to live or wear one's body a certain way," this process "only rarely becomes manifest to a reflective understanding."[26] Instead, we rely on recommendations from authority figures and friends to define how we act and who we are.

In a related move, Turing's depiction of AI is predicated on being able to perform normative gender roles, or "become a woman." Turing defined AI as the ability of a machine to appear human by predicting how one would answer basic questions about his or her gendered identity, a step in the direction of creating recommendations based on the personal answers given. Imagining computers as persons was an important move toward imagining them as personalizable. While this "imitation game" is well known, the heavily gendered aspects of the test and of Turing's conception of digital computers have largely been ignored. Other than Ed Finn's thoughtful description, the test is usually presented as follows: "An interrogator sitting in one room must ask two unseen participants questions to determine which is human and which is machine."[27] Yet Turing's original conception of the test actually starts with a man and woman each hidden from an interrogator in a separate room; the interrogator must ask them questions in order to determine their respective genders while the man pretends to be a woman. Turing then asks, "'What will happen when a machine takes the part of A [the man] in this game?' Will the interrogator decide wrongly as often when the game is played like this as he does when the game is played between a man and a woman?"[28] Here, Turing's model for AI is a transvestite that must prove its worth by passing not just as a human but as a man. In hindsight, given Turing's death after two years of estrogen injections that caused gynecomastia, among other things, this depiction of AI appears especially tragic.

Turing's imitation game should be viewed as a demonstration of Beauvoir's insight, as it illustrates how gender is an imitative performance that must be taught or, in this case, programmed. In the process, gender construction (and performativity more generally) becomes framed as an algorithmic process. Furthermore, the difference between human and machine was always already framed as an embodied performance analogous to the difference between genders. Removing the gendered aspects of this test makes it much easier to frame the history of AI and cybernetics as an "erasure of embodiment" rather than as a rethinking of embodiment focused on the transforming theories behind the relationship between gender, sex, and personhood.[29]

This queer image of computation is crucial for understanding how the logic of choice works within the deterministic framework of AI and recommendation systems in modern digital culture. Turing's test describes AI as only intelligent as long as it tries to pass for something else; otherwise, it is just artificial. Both Turing and Norbert Wiener discuss AI and digital computers as roughly analogous to humans, not because they both possess free will and the ability to make personal choices but rather because, in both cases, this sense of freedom is illusory.[30] Rather than being defined as a unique or individualistic experience (whatever that might be in this context), Turing's AI must conform to gender stereotypes and expectations; like us, it is preoccupied by a need to cast off its burden of choice and pretend that it is not free. Paradoxically, the more it does what we think it should, the more AI proves itself to be intelligent. Yet adopting the identity and life that one is supposed to adopt and appearing as one is supposed to appear, based on gender norms, capitalism, and other dominant ideologies, is a relinquishment of choice for humans and of intelligence for computers. Within such a framework, choice, subjectivity, and desire are only recognized as such when people or machines attempt to act as something they are not; choices are reduced to recommendations.

This logic of conflating choice with recommendations is key to contemporary culture and is perhaps most evident in the self-help industry. With its promise of self-enrichment (often both in a spiritual and a monetary sense), the recommendation is one of the central tropes of a self-help culture deeply tied to neoliberalism and postfeminism. Espoused by various culture industries and organizations, including cosmetics companies, support groups, dieting clubs, parenting guidebooks, and medical reference websites, self-help culture is both nebulous and deeply embedded throughout our contemporary mediascape. These industries use this formation to convert hegemonic commands into "advice" in ways that nevertheless offer a sense of autonomy and independence from institutions and authority figures. Advocating personal responsibility, self-management, and control, self-help industries encourage us to turn to and internalize vague and standardized advice meant for large populations rather than anyone in particular. Notably, there are not just countless self-help books but also innumerable recommendation lists for selecting those self-help books that are right for you. Thus, recommendations become a primary way in which this industry presents itself as personalizable, if not actually personal. This advice is meant not just to cure what ails its readers but also to make them better, fitter, and more productive workers, lovers, and consumers.

Yet, as Angela McRobbie argues, self-help industries create fears that people—and women in particular—will not be able to succeed in their personal lives.[31] These industries then sell a variety of self-management and self-monitoring routines, including cosmetics, exercise programs, dieting clubs, and continuing education courses, as the antidote to the fears they themselves created.[32] In *Tele-advising*, Mimi White details how the therapeutic and confessional discourses of self-help culture run through various television genres and construct the family and/as consumer culture.[33] White argues that television reconfigures the confession "as a public event, staged by the technological and signifying conventions of the television."[34]

I would further argue that recommendation systems play a similar role within digital media. As users reveal more and more personal information via social networks, search engines, shopping sites, and various apps for the promise of increased ease and guidance, digital technologies have only accelerated confessional practices geared toward figuring out both who you are in relation to others and who you could become; the recommendation is central to their continued appeal. It is perhaps not surprising that modern AI began with Eliza, a Rogerian therapy program designed to get users to reveal deeper and deeper truths about themselves in return for the promise of self-discovery. It is also not surprising that several websites that feature Eliza chatbots also now include personalized advertisements created by Google for, at least in my case, nearby psychologists.[35] Indeed, the technologies that allow for these ads and other digital forms of recommendation make users confess the personal (if perhaps no longer private) information stored in their browsers, computers, and online accounts related to their browser histories, locations, interests, and identities.

In their automated and algorithmic forms, recommendations now facilitate a whole host of contemporary neoliberal activities ranging from deregulation (the replacement of governmental edicts with privatized and unmonitored "recommendations") to global fair trade (by both making far-reaching products easier to find and flattening out their differences). At the same time, recommendation technologies' automated and seemingly objective nature makes them appear to ignore the gender, race, class, and other characteristics of users in ways that make all people appear equal through the eyes of the algorithms.

But like so much optimism, this belief that we are now color blind and live in an equitable world is simply a fantasy. Such problematic effects may appear quite familiar to anyone conversant with discussions of neoliberalism, postfeminism, and postracialism (or even those who have only glanced through their *Wikipedia* definitions). Recommendations now facilitate the growth of these discourses that paradoxically celebrate freedom while grappling with the value of personal choice. These ideologies spread not via orders but rather via recommendations—an enforcement mechanism designed to enact ideals of individuation and self-regulation by making subjects feel that the ultimate choice is in their hands. While these ideologies helped recommendations become a common interpellative technology, recommendations made these ideologies widespread.

Together, these ideologies argue (if they do not simply assume) that we have achieved a gender- and race-blind equality and therefore no longer have a need for feminist and civil rights activism. With such equality, they then argue, regulations are no longer necessary and the free market can now be truly "free." Indeed, I argue throughout this book that recommendation systems have a complex, but direct, relationship to this cultural tendency. On the one hand, these systems can manage some of the more damaging aspects of these trends by ostensibly making it easier to find what you want (whatever that might be) and quickly providing access to the same information and "best choices" as everyone else. On the other hand, they also facilitate these trends and thereby help to generate the very problems for which they provide solutions. They can make you more productive and efficient, which is what our capitalist economy demands, but they also make these demands appear reasonable when they are anything but.

Yet, even as recommendations attempt to appear natural, their fabricated and illusory dimensions are sometimes made visible—and therefore subject to critique. The quest for a perfect AI that is indistinguishable from a human neglects the possibility that the continual (and perhaps necessary) failure of AI leaves room for actual choice, for resistance. Recommendation systems and new technologies did not create neoliberalism, postfeminism, or postracialism (or any other ism), but they allow them to become more entrenched by reducing individual choice to consumer decisions. Real choice, however, moves beyond the recommended path. Many important feminist and civil rights milestones— from women's suffrage to *Roe v. Wade* and from the various civil and voting rights acts of the 1960s to the women's marches of the Trump era—empower women and minorities by broadening the choices they have in their own lives, including control over their own bodies, careers, and political aspirations. These victories showcase the political implications of preserving and broadening the scope of personal choice, privacy laws, and related cultural norms for everyone.

However, there has also been a continual pushback against expanding choices outside the commercial sphere. As Sarah Nilsen and Sarah E. Turner argue, throughout the 1980s, neoconservatives argued that the civil rights movement illustrated the strength of individual human agency in order to propose that all barriers to structural advancement for minorities and women had been eradicated and that therefore "any remaining racial inequalities were the product of individual choice rather than social policy."[36] Today, these discourses tend to frame choice as an expression of a specifically individualized freedom primarily performed in relation to the commercial marketplace (along with other important areas like charter school selections). While *Roe v. Wade* may encapsulate the meaning of choice within the women's rights movement, post-feminism too often defines choice in terms of a woman's power to buy what she likes. Likewise, neoliberalism defines freedom in terms of the ability of the individual to participate in the free marketplace.

Yet, at the same time, these same discourses largely frame the choices of minorities as a problem. Imani Perry argues that the American legal system frames being "poor and black or brown" as a "bad" choice in order to justify the outrageous level of surveillance that surrounds and invades these communities.[37] Furthermore, Catherine R. Squires argues that "post-racial discourses obfuscate institutional racism and blame continuing racial inequalities on individuals who make poor choices for themselves or their families."[38] Indeed, for Squires, the logic of the recommendation drives postracialism by creating an assimilationist bias that frames the ideologically dominant (white) choice as the only real possibility. Within this framework, recommendations become indistinguishable from commands.

This debate illustrates how the contours of choice are always culturally situated and politically motivated in ways that reveal dominant ideologies and power dynamics. As Rosalind C. Gill has argued, certain choices, especially those made by women and girls, "are not respected as choices but are repeatedly understood in terms of external influence (from religion, from consumer culture) rather than as authentic, autonomous acts."[39] Gill poses the question, "What kind of feminist *politics* follows from a position in which all behavior (even following fashion) is understood within a discourse of free choice and autonomy?"[40] Like her, I am particularly concerned with how contemporary culture uses discourses of choice and recommendation to invite women to "become a particular kind of self . . . endowed with agency on condition that . . . it is used to construct oneself as a subject closely resembling a heterosexual male fantasy."[41] Recommendations have become a modern tool to sell this postfeminist fantasy of "the possibility that women might *choose* to retreat from the public world of work" and instead gain fulfillment by primarily supporting their husbands' careers instead.[42]

For instance, in 2014, *Cosmopolitan* published a controversial article on a woman who "chose to be a stay-at-home wife."[43] Yet, as the article makes clear, her workload had tripled and her salary was stagnant, and her husband was working a sixty-hour-a-week job as well. As she describes washing her dog and doing housework, she explains that she used to hire people to do this work for more money than she made herself. By framing her decision to leave her job as a personal choice, this article obfuscates the economic necessities that drove her to her decision and instead becomes a recommendation for this way of life, regardless of its economic impact. Examples like this illustrate why we must be wary of how various groups employ discourses of choice and recommendations with both feminist and postfeminist aims. As one side lauds personal agency and choice, the other side proposes the need for more regulations and recommendations (and vice versa).

GroupLens and the "Benign" Algorithmic Solution

The World Wide Web has often been presented as an attempt to create a more equitable world by (at least theoretically) expanding the choices and opportunities of many by making it less necessary for them to divulge their gender, race, sexuality, and identity to those with whom they come into contact. Yet much of the history of the World Wide Web has been focused not on expanding choice but rather on solving the dilemma of having too much of it. From nearly the very beginning of the World Wide Web, computer engineers envisioned recommendation systems and the collaborative filtering algorithms they contain as the answer to the dilemma of having too much choice—a problem that few had yet encountered online. In the early 1990s there were several academic groups and private companies across the globe creating recommendation systems. While in the next chapter I focus on the early work of Pattie Maes and her students at the Massachusetts Institute of Technology, here I discuss the efforts of the contemporaneous GroupLens Research Lab at the University of Minnesota, which started working on collaborative filtering in 1992. Headed by computer scientists John Riedl and Paul Resnick (and, soon thereafter, Joseph Konstan), GroupLens created technologies to make it easier for people to find interesting and relevant information online. Their first project, the GroupLens Usenet Article Recommender, used collaborative filters to locate and recommend Usenet news (or Netnews) articles. Similar to an early internet forum or bulletin board system, Usenet was an early dial-up computer network wherein users could read and post articles sorted into various categories. Posted articles first went to moderators who could decide which category they should be placed in.

While these categories were helpful for arranging and locating articles, by 1994 there were more than eight thousand newsgroups, and one study showed

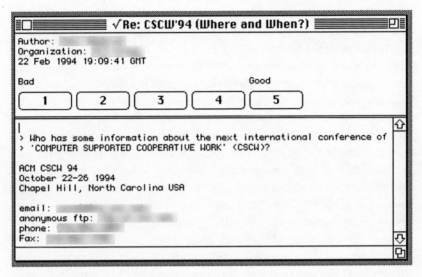

FIG. 1.1 Screenshot of GroupLens program, 1994

that 140,000 people posted stories over a two-week period.[44] In a 1994 Association for Computing Machinery conference paper explaining GroupLens, Resnick and his coauthors argued that, while better than nothing, this vast number of articles made Netnews "not completely satisfying," as it was hard to figure out which articles were worth reading and whether your comments on stories were valued: "Each reader ends up sifting through many news articles to find a few valuable ones. Often, readers find the process too frustrating and stop reading netnews altogether."[45] They argued that making a decision when there are seemingly endless choices, even in this extremely low-stakes environment, is simply too difficult; many would apparently rather just not make a choice at all than make the wrong one.

First, the authors defined the large number of articles on Netnews as a problem within an increasingly capitalistic online culture focused on efficiency above all else; then they created and eventually sold their own solution to their problem. In response to these issues, Riedl and colleagues created Group-Lens, a program that allowed users to rate the articles they read and get automatically recommended articles that others with the same taste liked in the past. They framed the collaborative filter as a "deceptively simple idea" that could change not just how people get online content but also how they socially interact online.[46] They argued that with collaborative filtering, the value of moderation, and the human authority that comes with it, would decline. This filtering, which resulted in every user's seeing a unique list of articles, could make any standardized categorization much less influential, and the authors questioned whether this would result in a more socialist-styled system of permeable "peer groups with shared interests" that were willing to work with each other to make the community better or whether it would instead fracture the "global village" into tribes.[47] Eli Pariser and others argue that many are still quite anxious about whether recommendation systems limit our world-view by making it so that we only receive our news and media from those with a similar perspective.[48] While such recommendations may make reading the news more efficient and perhaps more pleasurable, they may also narrow our interests and isolate us from others who do not share our interests.

The fracturing of audiences based on interests and desires is not a new problem but is rather a central aspect of the experience of giving and receiv-ing recommendations. In an interview for this book, Konstan, who has been working with GroupLens since 1993 and taught a massive open online course on recommendation systems, explained that the GroupLens team and many others at the time felt that there were simply "too many choices" online and that bringing the logic of recommendations to online spaces was an elegant solution.[49] At the same time, he noted that much of their work in the mid-1990s was inspired by the idea that the economy was becoming more focused on the value of information. They saw that there was a great deal of online user activity (in the form of, among other things, clicks, reading habits, and prefer-ences) that at that point was simply disappearing, and he and his colleagues felt it was a "no-brainer" to make this information both valuable and useful for users and companies alike; recommendation systems helped make this "infor-mation economy" possible. Konstan pointed out that while GroupLens played an important role in experimenting with recommendation systems, there were many other labs and companies filing patents and working on similar issues. In 1996, over forty-five recommendation system projects were dis-cussed at the first workshop on collaborative computing at the University of California, Berkeley.[50] At the time, the sheer number of groups working on this technology made the value of recommendation systems appear obvious and their dominance seem inevitable.

Yet in creating and applying these technologies, Konstan was also quite aware of how subjective they were. Recommendation systems did not create a more natural or necessarily better way of organizing information. While these technologies appeared to make the act of choosing more efficient, he compared their effect to that of putting candy rather than fruit at the checkout stand: it may create more sales, but that does not mean it is better for you.[51] This anal-ogy illustrates that recommendations became a useful way to organize online data not because they were simply the best or most natural form but rather because they fit in with and facilitated the growth of the capitalist informa-tion economy, which benefits the most affluent among us.

Over time, the utility and profitability of recommendation systems became apparent across the e-commerce industry. In 1996 a handful of GroupLens professors and students, including Konstan, Resnick, and Riedl, started Net Perceptions, a company that sold custom-built recommendation systems to various companies. In the fall of that year, they sold Amazon a recommendation system that has since morphed into one of the most visible and influential uses of this technology. They also sold systems to various other major companies trying to start e-commerce sites, including 3M, Best Buy, Kmart, JCPenney, and (my personal favorite) the now-defunct SkyMall.[52] At the same time, the GroupLens lab started adapting its technologies to other uses and created MovieLens, a movie recommendation system that continues to be widely used by the public and experimented on by academics.

During the late 1990s Net Perceptions, and recommendation systems more generally, were viewed as a utopian answer to many of the early cultural anxieties about digital technologies. In *Word of Mouse*, a trade book that both explains and glorifies Net Perceptions and collaborative filtering technologies more generally, Riedl and Konstan argue that this technology can allow marketers and others to focus on individuals rather than groups as a way to "box products, not people."[53] They felt that recommendations could empower users while making companies rich.

At the same time, this focus on individuals raised new fears of exploitation and anxieties that these technologies were merely re-creating stereotypes under a new guise. If the GroupLens lab originally voiced fears that recommendations would split us into tribes rather than gather us into a global village, by 1999 these fears had been inverted. After a decade that saw publishing and film industries completely dominated by global blockbusters, the cultivation of a single, pervasive monoculture rather than insular microcultures became the threat. In a 1999 *New Yorker* article on recommendation systems, Malcolm Gladwell calls collaborative filtering "the anti-blockbuster."[54] He explains that he rarely sees anything but commercial Hollywood releases and was impressed that after he rated a few of these films, MovieLens recommended an obscure Belgian comedy and a 1937 Fred Astaire and Ginger Rogers musical. By recommending obscure films based only on an interest in certain blockbusters, Gladwell argues, collaborative filtering could reshape not just the film industry but also the "book market" by making customers aware of new or lesser-known authors that they specifically may enjoy.

Suddenly, recommendations made tribalism the solution rather than the problem. While tribalism had been associated with racism and xenophobia, collaborative filtering seemed to promise a way to define people by their tastes without seemingly taking race, gender, class, or other identity markers into account. Gladwell points out that while collaborative filtering splits people up into "'cultural' neighborhoods," which Wendy Hui Kyong Chun argues echo traditional marketing's reliance on harmful and inaccurate stereotypes, these new practices create stereotypes that are not static but rather continually transform from one click to the next.[55] These recommendations appear to separate taste from culture—one's "cultural" neighborhood from one's actual neighbors. In the process, they facilitate the postracial and postfeminist fantasy that race, gender, and sexuality are no longer meaningful ways of defining and sorting people, online or off. Konstan and Riedl indicate that rather than lumping us into cells, collaborative filtering puts us in a "melting pot. We're black inner-city youth, upper-class middle-aged WASP, conservative southern landowner, working poor, etc."[56] While they argue that "these boundaries are increasingly difficult to extricate ourselves from, no matter how much we protest," recommendation systems offer a potential escape route.[57] Yet, while the melting pot image is perhaps better than that of cells, it proposes that these technologies do not just ignore but actively erase ethnicity from users even as they are treated as "individuals." Paradoxically, Chun has also argued that this technology and others like it use taste as a proxy to "reinforce categories such as race, gender, [and] sexuality."[58] If nothing else, these two disparate perspectives illustrate the continuing tension and confusion over the assumptions and goals of the industries that employ these technologies. Chun does not give any specific evidence or reasoning for how this reinforcement works, but it is clear that, like postracialism writ large, much of the hype and excitement around recommendations rests on a separation between culture and taste that has never been more than a fantasy.

Eventually during the dot-com bubble, Net Perceptions went public and was briefly valued at a billion dollars. Then, after the industry-wide crash, its sales became stagnant, the company stopped growing, and, in 2004, it was liquidated and sold off in pieces.[59] While Net Perceptions was just one of many companies during this period that focused on recommendation systems (and I will discuss another, Agents, in the next chapter), its trajectory reflects larger trends throughout the industry. Konstan argues that Net Perceptions eventually failed partly because recommendation systems succeeded. As recommendations became a central mode of marketing, website design, and personalization, companies began to want them to be a core part of their technological infrastructure.[60] As Net Perceptions only sold recommendation systems, rather than a whole suite of personalization and e-commerce services, it became less desirable than those that provided the full range of such services. The more important recommendations became to e-commerce, the less able Net Perceptions was to capitalize on it.

Yet Riedl and Konstan also clearly envisioned how important not just collaborative filtering but recommendations more generally would be for the future of the web. In *Word of Mouse* (2002), they declared that collaborative filtering "is at the same time very new and very old" and linked it to prehistoric forms of recommendations that kept cavemen from eating the wrong foods and from approaching the wrong animals.[61] While not a matter of life or death, collaborative filters do function to create communities between users and connections between people and companies in the way recommendations always have, but Riedl and Konstan do not point out that, just as often, recommendations pull people apart and destabilize communities by defining what is good, what is normal, and what is not. Riedl and Konstan assert that, due to television and other forms of "mindless entertainment," the very idea of community was in decline, but recommendations could reinvigorate "the desire for human interaction." Online, recommendations encourage people to "make connections that they never would have before. People in small Wyoming towns can meet people in Madrid who share their interests."[62] Much of *Word of Mouse* focuses on how to make these connections stronger and more positive by, among other things, bringing users into contact with a company's brand or store more often and by backing up the automated recommendations with expert knowledge. Since recommendations say as much about the recommender as they do about the product, Riedl and Konstan advise that companies do everything they can to make their recommendations appear thoughtful and caring.

For better and worse, *Word of Mouse* prophetically describes the future of the web and puts recommendations at the center of the transformations they predicted it would undergo. Its authors envisioned "colossal battles" concerning how recommendations could be used and who would control and own all the personal information and data that these technologies record.[63] The authors foresaw how recommendations would one day be used not only to sell content to users but also to have users create content and even turn user communities into content (as happens on social networks like Facebook). They detail the role recommendations would play in the creation of valuable sticky content (as on news sites like Buzzfeed) and viral marketing. As they document, many companies were experimenting with these uses in the late 1990s and early 2000s, far before the web 2.0 rush, but it is surprising to see how many of these efforts became mainstream and common in the years ahead. The authors even foresaw the role recommendations have only now begun to play in augmented reality and smartphone applications as a tool to acquaint users with and connect them to the space around them.

While many of these early companies disappeared, they were replaced by many more that now predominantly sell recommendation systems as part of larger e-commerce packages that also include various forms of online marketing and website personalization services. These services facilitate the recent normalization and massive scale of the recommendation industry itself. Now, Barilliance, Baynote, and Certona, among many other companies, sell generic recommendation systems as part of a package of personalization technologies to hundreds of companies, from multinational corporations like 3M, LG, and Urban Decay to more specialized and localized merchants like Acuista and FotoKasten. Illustrating the role of recommendation systems in shaping not just popular but also high culture, Certona provides its software to New York's Museum of Modern Art online store and in 2015 was responsible for 12 percent of the museum's sales.[64] These companies design their recommendation systems to work easily with any product or service; Barilliance advertises that it can integrate its system into any online store in only five minutes and promotes that its recommendations have improved sales rates and conversation rates for clients by as much as 500 percent.[65] These companies lease their recommendation systems in packages with a host of related "personalization" technologies. They advertise these packages as a way to make it more possible for users to experience freedom by making it easier for them to make their own choices. But at the same time, they constrain users and turn them into the products themselves.

Now, these companies present digital recommendation systems as a natural part of the digital landscape. The more omnipresent they become, the less visible they are. Barilliance, Baynote, and Certona present recommendations as a way to improve website search tools, transform the main pages of sites to feature products that individual users would be more likely to purchase, and send these users personalized emails based on their recommendations.[66] The ubiquity of these systems not only transforms the way the consumer web works for everyone but also quietly structures every choice made online as a recommendation. The companies that use these technologies do so in order to create desires, dreams, and needs that drive subjects not just toward more consumption but also toward particular patterns of consumption under the idealized guise of free choice.

Relieving Us of Choice Online

As these discourses and technologies pertaining to choice and recommendation construct consumer decisions as an alias for choice, such choice becomes increasingly reserved for the affluent. David Harvey describes how "the idea of freedom 'thus degenerates into a mere advocacy of free enterprise,' which means 'the fullness of freedom for those whose income, leisure and security need no enhancing, and a mere pittance of liberty for the people, who may in vain attempt to make use of their democratic rights to gain shelter from the power of the owners of property.'"[67] The examples of GroupLens and other companies like it illustrate how discourses about freedom and choice focus solely on the affluent elite, which "tends to confuse self-interest with individuality and elevates consumption as a strategy for healing those dissatisfactions that might alternatively be understood in terms of social ills and discontents."[68] The valuing of an extremely restricted sense of free choice supports this substitution, first, of personal liberty for freedom and, second, of free enterprise for personal liberty. Neoliberalism makes it clear that individuals "are not supposed to choose to construct strong collective institutions (such as trade unions) as opposed to weak voluntary associations (like charitable organizations). They most certainly should not choose to associate with or create political parties with the aim of forcing the state to intervene in or eliminate the market."[69] Harvey demonstrates that free choice is typically only celebrated as long as it leads to the "right," recommended decision.

Indeed, this movement toward conservative choices—those having a stabilizing and positive effect on the market—is achieved not through brute force but rather through the ideological logic of the recommendation—and the technology of the recommendation system. Recommendations connote authority and are supposed to propose the "best" choice possible. As such, they indicate not a free choice but rather one circumscribed by larger structures of power and control. "Real" choices become those made in opposition to recommendations—as those choices that fall "outside the box," "take the road less traveled," and are otherwise not regarded as a safe route to happiness in the guise of economic success.

Renata Salecl has argued that accepting the decisions and recommendations of others relieves the anxiety and even trauma of making choices for ourselves; it is the way "we choose the option of not choosing—of having our choices made for us."[70] Recommendations offer relative safety and stability and, in so doing, present real choice as an unsafe, potentially costly burden. Indeed, the struggle and failure to be free is central to how a liberal sense of choice as a generator of free will is discursively produced and performed by subjects. Digital recommendation technologies rely heavily on information-gathering and surveillance techniques that limit user privacy in online spaces as a way to not increase our freedom but rather relieve us of it. Moreover, as this surveillance constructs recommendations, it also curbs the agency of users in making choices.

Online advertisements and self-help regimes do not just instrumentalize the recommendation but in fact turn the recommendation itself into a commercial product sold from one business to another. Companies rely on recommendation's connotations of faith and personal connection to present themselves as benevolent and caring when they are anything but. Such companies transform recommendations into what Sartre calls "bad faith." He describes acts of bad faith as those many strategies we use to try to pretend or convince ourselves that we are not actually free. For example, he describes a waiter who moves and acts exactly as you would expect a waiter to, but more so: "His movement is quick and forward, a little too precise, a little too rapid."[71] Rather than describe him simply as a very good waiter, Sartre argues that "he is playing, he is amusing himself. But what is he playing? We need not watch long before we can explain it: he is playing *at being* a café waiter."[72] Sartre complains about this hyperperformance because it seems to simply be a copy of what he has seen other waiters do and has lost its authentic individuality. On the one hand, simply repeating this model becomes a safe route to success as a waiter. But on the other hand, Sartre argues that the desire to fit into that model is ultimately self-deceptive; the adoption of such a model becomes a way for us to deny our freedom to be whatever we want; to be ourselves.

The companies that turn recommendations into advertisements actively monetize this bad faith by presenting their recommendations as a relief from having to make a choice. Indeed, recommendation systems are designed to manage and alleviate this burden by granting certain choices a seal of approval, thus making them appear both more natural and more compulsory. Companies that create and use recommendation systems are responding to the fear that too much choice can actually lessen profits. Applying Sartre's "burden of freedom" to business policy research, economists Sheena Iyengar, Mark Lepper, and Barry Schwartz have referred to this phenomenon as the "paradox of choice."[73] In his book of the same name, Schwartz argues that too much choice creates anxiety and makes it more difficult to come to any final decision based on any qualitative rationale.

Experiences of abundance and heterogeneity become channeled into the "recommended choice." Salecl has further argued that anxiety about choice is compounded by "today's consumer society," in which "we are asked to see our whole lives as one big composite of decisions and choices."[74] Like Sartre, marketing scholars and pundits understand that freedom is a burden, but not one for which individuals should be responsible. Rather, they illustrate and monetize the pleasures that can come from limitation. Instead of freedom, the recommendation industry represents this abundance of contemporary choices as a "path of self-destruction" that Salecl somewhat hyperbolically links to contemporary epidemics of "self-harm, anorexia, bulimia and addictions."[75] As a response to this "crisis," recommendation systems point to the larger socioeconomic ideologies that work to structure subjectivity.

By encouraging us to follow recommendations, primarily by purchasing products, these services sell us the opportunity to pretend that we are not free and that we are not ultimately the ones making the choice. Many popular websites, including Google, Netflix, and Amazon, employ recommendation systems to assist users in making preset decisions of all types in ways that echo Max Horkheimer and Theodor Adorno's concept of the "culture industry." Whereas Sartre described how people ultimately and knowingly are the arbiters of their decisions, Horkheimer and Adorno described how mass industries do everything they can to delude us into thinking we are making our own choices when we are really only following their recommendations. They generate various standardized products that are only superficially distinct. Deciding between such products may feel like a type of freedom, but such products are still limiting and become a way to deny ourselves a greater type of freedom. As Horkheimer and Adorno describe, these superficial differentiations "do not so much reflect real differences as assist in the classification, organization, and identification of consumers."[76] This logic is central to the present digital economy, where recommendations and search results (which are roughly the same thing) get stored and analyzed in order to define users more accurately and better relieve them of their burden of choices over not just what to purchase but also what to become.

While they rarely produce standardized goods, those many companies that employ recommendations now constitute a culture industry that generates standardized shopping choices and life decisions. This industry uses recommendations in ways that generally uphold the status quo and steer mass culture toward greater conformity. They turn choices and recommendations into technological and ideological instruments. Whether in the form of Google's or Bing's organization of search results and presentation of personalized advertisements, or that of Pandora's or iTunes Radio's automatically generated music playlists, these industries use recommendation systems to lead users toward certain information, objects, entertainment, purchases, and decisions and away from others. Their automated recommendations, derived from an analysis of user data, shape the contemporary self through a rhetoric that equates conformity with equality and consumerism with freedom. In the process, the digital recommendation system has become a primary instrument of control that encourages users to think of increasing choice as a problem and recommendations as the "benign" solution.

Resisting Recommendations

Riedl and Konstan end *Word of Mouse* with a description of a day in the life of a marketer in the not-too-distant future. Every one to two hours throughout the day, the marketer uses recommendations to make business decisions and purchase personal goods and gifts. After dinner, he "toasts the brave new world with a scotch. It's a single malt he's never tried before—but one he's sure he'll like."[77] In what appears to be an extreme misreading of Aldous Huxley's classic, recommendations become not just a normal but a positively utopic tool for the creation and maintenance of capitalism and its sedate subjects. If nothing else, this image exemplifies how recommendations serve both as a simple, pleasurable convenience and also as a form of control, as key elements of a "golden era of marketing" that will usher in a brave new world, and as a tool to alleviate our burden of freedom. Much like soma itself, recommendations constrain us through the promise of pleasure, or at least the absence of bad choices and the pain they may bring. But when you get rid of bad choices, what is left?

This chapter has charted the long history of how recommendations have been used not to increase but rather to circumscribe our freedom of choice in sometimes extremely lucrative ways. In this context and in the face of these forces, making a choice rather than following a recommendation can be represented as a downright subversive expression of self; as such choices are not necessarily a direct challenge to the system of recommendations themselves, they do require that one go against the grain and become aware of the contours of dominant ideology and of common sense as an artifice. By making an unrecommended, or "bad," choice, we reclaim our burden of freedom. As in AI studies, choice in various sectors of the civil and women's rights movements is framed as transgressively gendered and racialized, as what Judith Halberstam calls the "queer art of failure." Halberstam argues that "under certain circumstances failing, losing, forgetting, unmaking, undoing, unbecoming, not knowing may in fact offer more creative, more cooperative, more surprising ways of being in the world."[78] She outlines how feminism has long valued failure within heteronormativity as it allows people a chance to productively rethink and remake themselves, their communities, and their gendered identities.[79] When applied to "the liberal concept of choice," queer failure becomes the act of purposefully making "bad" choices as a way to challenge and devalue those choices recommended and considered "right" under the capitalist, heteronormative logic of liberalism.[80]

Far too often, choices become perceptible as such only when they fail to achieve a desired goal or are made in opposition to a dominant ideology (i.e., that which is recommended). As both Sara Ahmed and Jasbir K. Puar stress, choices that achieve their goals or are made in agreement with a dominant ideology are discursively framed simply as compulsory, as imperious, as common sense.[81] Kara Keeling argues, common sense provides, "in the form of clichés, a way of continuing present movements."[82] Through the proxy of recommendations, common sense habituates us to, and rationalizes, various "forms of domination and exploitation."[83] In contrast, this logic constitutes unrecommended choices as subversive non-sense that points to forms of success and happiness that are not authorized by popular culture.

Ahmed and Puar have demonstrated how feminism and neoliberalism are awkwardly aligned in representing those choices that simply follow the dominant ideology as not choices at all. Queer subjects come to be defined by their transgression of all given choices rather than by an identity that is not a choice but perhaps driven by biology or another compulsion. As Ahmed and Puar note, defining queerness as a freedom from norms ironically limits this subject position to elite cosmopolitans with social mobility, capital, and resources.[84] It also limits definitions of agency to only those choices and actions that are resistant to the status quo. Yet our relationship to recommendations is far from one of simple, passive conformity; just because we are only given a set number of choices from which to select or recommendations designed to show us how to be more like others (who are, already, like us) does not mean that we cannot question these recommendations and use them in our own playful and nefarious ways. In the following chapters, I examine instances in which confrontations with recommendation systems did not go smoothly and pushed users to question, play with, and at times challenge the relationship between choices and recommendations throughout digital culture.

2

Female Labor and
Digital Media

• •

Pattie Maes and the Birth
of Recommendation Systems
and Social Networking
Technologies

Histories of the beginning of the internet, the World Wide Web, and digital culture too often leave women and minorities out entirely.[1] While theorists and critics like N. Katherine Hayles, Wendy Hui Kyong Chun, and Sadie Plant have brought to light the central role of women in the creation and use of many of the first computers, popular histories on the emergence of social networking and contemporary web culture like David Fincher's *Social Network*, Steven Levy's *In the Plex: How Google Thinks, Works and Shapes Our Lives*, and Eli Pariser's *Filter Bubble* often frame these technologies and practices as having been created overwhelmingly by white men. It is hard to tell from these texts whether their authors actively obscure the contributions of women and minorities who did play a role in the creation of these technologies or whether their absence in these texts reflects a lack of sufficient research. However, the effect is clear: as Anne Marie Balsamo has argued, the declaration that it is "the class of white men who have enjoyed the benefits of formal institutional recognition as agents of the technological imagination" systematically writes women out of "the historical record of technology development."[2]

While in the last chapter I discussed the male-dominated GroupLens Research Lab in relation to the long cultural history of recommendations, here I focus on the 1990s work of Massachusetts Institute of Technology (MIT) professor Pattie Maes, one of the most visible and influential people working on recommendation technologies.[3] Twenty years after the feminist revolution, white male faculty dominated MIT. During this period, as women entered more and more executive and professional workplaces, Maes created software to address the many difficulties she faced in balancing her own academic and family life. Working within a patriarchal framework, she authored a number of groundbreaking programs during that period: one that could automatically schedule meetings for users, one for prioritizing and sorting emails, one of the first matchmaking programs, and one of the first online social networks. While she never achieved the same fame as her colleagues Nicholas Negroponte and Sherry Turkle, her work has been just as, if not more, influential in the development of digital culture.[4] Her programs served as prototypes for the infrastructure of the World Wide Web, including digital recommendation systems, password management and user identity protections in online consumerism, and the culture industries of choice that these technologies support. These industries guide users toward certain information, products, and services over others with a standardized method of presenting choices and recommendations to users. Together with those of GroupLens and various other labs, Maes's technologies led to the industrialization and instrumentalization of choice.

Even as these technologies facilitated certain neoliberal, postfeminist, and postracial ideologies, Maes wanted her work to confront gendered problems involving labor, time, and self-management. By focusing on the work of Maes—instead of starting ten years later with the birth of Google, YouTube, or Facebook—a very different history of social networking and digital culture emerges that brings gender and race into the foreground.

The Postfeminist Subject in the Digital Era

To better understand these recommendation and social media technologies, it is important to address how the postfeminist context in which they were created helped to shape the ways they are currently used. Specifically, this period saw a pushback against changes in gender norms around family and work, which played an important role in the history of social media. I argue here that postfeminist discourses on choice in relation to family-work dynamics and difficulties played were crucial in laying the ideological framework for digital recommendation systems. Indeed, as I discussed in chapter 1, neoliberal and nascent postracial discourses also played important roles, and these ideological forces cannot be separated. Consumer industries have tended to value affluent white women more than minorities and the impoverished because these women are perceived to be more valuable customers. This leads to marketers encouraging other cultural groups to conform to this privileged identity. For this reason, of these three discourses, postfeminism tends to dominate in these industries, which makes minorities and the poor all the more invisible.

Postfeminism in the 1990s, as Diane Negra and Yvonne Tasker have argued, first and foremost presupposed the "'pastness' of feminism, whether that supposed pastness is merely noted, mourned, or celebrated."[5] At times, scholars and journalists have framed this period as being characterized by a backlash against the victories of feminist fights for gender equality, and at other times they have depicted it as one in which those victories were taken into account.[6] While feminism focuses on creating gender equality in the workplace and all other areas of life, postfeminism instead celebrates the liberatory potential of personal choice within women's lives. Postfeminism asserts that gender equality has been achieved and that this equality is expressed through both the life-changing and the everyday choices that women are now free to make. Negra has argued that by focusing specifically on choice, postfeminism replaces a discourse on equality and communal need with one that privileges individuality and personal desire, especially as they are expressed through consumerism.

Postfeminism's celebration of the subject as an individual is deeply connected to the neoliberal focus on personal responsibility and free trade, which depends on everyone behaving, regardless of financial means, as if they are in charge of their own destinies. As Angela McRobbie, Lisa Duggan, and others have argued, neoliberalism, postfeminism, and postracialism are intertwined, and all seek to replicate (and forcefully imagine) a system of unmarked individuality by insisting that sexism and racism no longer exist and that everyone now agrees that there are no socially significant differences among genders, races, and ethnicities.[7] Neoliberalism only works as a guiding socioeconomic structure if we all continually self-manage and behave as if we are all autonomous agents operating on a level playing field.

However, women (and especially women in the United States) are the ones to whom self-discipline and self-management are marketed most heavily, as attested to by the entire beauty and self-help industries aimed primarily at the female consumer. At the same time that postfeminism celebrates the liberty and equality of women, it often recommends those choices that reinstate the normative patriarchal gender roles that feminism worked to deconstruct. For example, in a 2013 *CBS This Morning* roundtable discussion on the state of contemporary feminism, *New York* magazine journalist Lisa Miller argued that feminists should have the choice not to work and instead devote themselves to becoming better housewives because they are not naturally as ambitious and competitive as men. Relying on conservative essentialist logic, Miller asserted, "mothers instinctively want to devote themselves to home more than fathers do," and feminist efforts to close the wage gap and bring more women into the workplace are misguided and harmful.[8] Postfeminism's focus not just on choice but on the biologically correct choice as the key to independence and happiness and its assumption that equality has been achieved neglect and obscure the continued pay differentials between men, women, and minorities, as well as the many other ways in which women continue to be disempowered in the workplace.

Postfeminist logic also participates in neoliberal efforts to keep many poor and undereducated women out of the workforce "because the sort of employment they're most likely to obtain won't cover the costs of childcare."[9] Furthermore, postfeminism construes feminism's hard-won rights as a burden that causes women anxiety and stress; through a rhetoric that celebrates the liberatory potential of certain choices (i.e., recommendations), postfeminism can argue that taking on traditional gender roles and becoming a housewife is a preferable lifestyle. As Negra has argued, postfeminist culture continually reinforces "conservative norms as the ultimate 'best choices' in women's lives."[10] By asserting that the desire to work is a problem that feminism generated, postfeminism abandons both the fight for equal pay and rights and the vast majority of women for whom work is not a choice or recommendation but rather a necessity.

While postfeminism celebrates the housewife, it also suggests that women should be able to work in any field but encourages them to think of these jobs as a way to improve their families and lives outside work. In this vein, postfeminism focuses on the affluent, overwhelmingly white, female professional for whom work is a choice. This focus frames the "privilege" to work and spend money as the definition of personal freedom for women who either are free of family obligations or can use their income to ease their home workload by hiring a cleaning service or a nanny. While feminist social movements altered gender roles by fighting for gender equity and rights, postfeminism uses a rhetoric of pastness that argues that women are now "free"; this temporizing rhetoric transforms equity and rights into freedom, privilege, and individuation through choice. The valorizing of the female professional's "freedom" to make money in order to consume excludes socioeconomically disadvantaged women and, as Tasker and Negra put it, "commodifies feminism via the figure of woman as empowered consumer."[11] Moreover, while this new ability of a woman to choose how she lives her life is celebrated, it is also seen as a burden. While women are told that they can "have it all," this slogan is often translated into the burden of "doing it all," as they are still expected to manage their families even as they gain higher positions and more responsibility in the workplace.

Postfeminism not only transformed equality into the freedom to consume; it also framed that freedom as a problem needing solutions. This logic asserts that the sheer number of choices and the need to make them on a continual basis have resulted in choice becoming, at least at times, a burden rather than a freedom. Indeed, Negra has pointed out that rather than leading to more liberty, postfeminism supports "distorted renditions of choice" that is too often oppressive.[12] Rhetoric concerning the need to make choices presents them as a source of anxiety filled with the possibility of making the wrong ones and ending up alone in "emotional isolation" with no one to blame but oneself.[13] Indeed, the postfeminist era is full of new choices for affluent women who frequently pursue the goal of having a family while both parents work as professionals.[14]

This postfeminist situation applies not only to women but also to men. Both men and women in the professional elite must now make a host of choices— concerning everything from major decisions like whether and when to have kids, to the need to arrange their daily schedule around professional and famil- ial commitments. Furthermore, postfeminist culture frames the "self," male or female, as a project, never finished, which takes up all of one's time. Never- theless, women's lives especially "are regularly conceived of as time starved; women themselves are portrayed as overworked, rushed, harassed, subject to their 'biological clocks,' and so on to such a degree that female adulthood is defined as a state of chronic temporal crisis."[15] As more women entered profes- sional careers in the 1980s and 1990s, time pressures increased; these women had to divide their time between their careers and families and decide which to prioritize.

Postfeminism presents time as a scarce resource divided among work, rela- tionships, leisure, and especially personal development. While being "time starved" and "overworked" affects people of all classes, genders, and races, McRobbie argues that a bevy of advertisements, texts, and films problemati- cally narrativize these anxieties as a specifically female and overwhelmingly white problem brought on by white women's ever-increasing presence in the workplace.[16] Many men also struggle with managing their time between work and home life but are not encouraged to engage in personal development to any- where near the same degree as women. However, as there were 12.4 million single-mother households in 1994 (compared to 2.9 million single-father households), it is clear that the difficulties and stresses of managing work and family were, in the 1990s at least, a gendered problem that affected women much more than men.[17]

Central to this temporal crisis is postfeminism's paired focus on self- discipline and self-management via the self-help industry. This focus on female self-management is intertwined not just with the current cultures of postfem- inism and neoliberalism but also with our current technological imagination and the longer history of femininity in the West. Rosalind Gill and Christina Scharff argue that the ideal neoliberal subject is actually a woman, or at least a postfeminist woman or girl. Even as neoliberalism and postracialism assume a subject that does not believe itself to be exposed "to pressures, constraints or influences from outside," women and minorities historically are called on to self-manage and self-discipline in ways that white men are not. Gill and Scharff state, "To a much greater extent than men, women are required to work on and transform the self, to regulate every aspect of their conduct, and to present all their actions as freely chosen. Could it be that neoliberalism is always already gendered, and that women are constructed as its ideal subjects?"[18] I think so, with white bourgeoise women being the most ideal. However, just like the first- and second-wave feminist movements, postfeminist, postracial rhetoric and neoliberalism affect both women and men of all races in a variety of overlapping ways. As Steve Cohan has argued, when men attempt to self-discipline through physical or behavioral change or control, they are often represented negatively as feminized (neoliberal, postfeminist) subjects.[19] Women are supposed to self-manage. Men are not supposed to need to—though they are actually encouraged to do so in myriad ways.

This drive to self-discipline through consumerism has grown alongside the expansion of consumer culture in America throughout the 1990s and 2000s. While in 1994 there were five hundred thousand consumer goods available in America, by 2003 there were nearly seven hundred thousand.[20] Although many of these products are sold in stores (with the average supermarket carrying around forty-five thousand items and big-box stores like Costco often carrying over one hundred thousand items), the bulk of them can only be found online on sites like Amazon and Alibaba, where the physical constraints of location often seem to matter less.[21] As Sheena Iyengar has argued, this increase in consumer choice, which often is equated with personal freedom, causes individuals to "give up the best option(s) for a wider range of inferior options" and leads to feelings of anxiety, guilt, and a diminished enjoyment in any choice; the availability of too many options makes people worse at making a decision.[22] The promise of postfeminism—that making consumer choices "will allow us to 'realize' ourselves, to be and have all that we ought to be and have"—is "turned back on to us."[23] Postfeminism demands that people make consumer choices to discover and define not only themselves but also, and more importantly, what a "self" can be. And digital technologies have been actively designed to assist people in crafting these definitions. Ideally, in digital spaces, the user is the self. However, recommendation technologies frame the user and hence the self as a consumer; in the process, using and being become aliases for consuming.

Autonomous Agents and Postfeminist, Digital Subjectivity

While I argue that postfeminism and other, related individualistic ideologies shaped digital technologies, it is also clear that technology transformed them as well. Maes's biography and self-representational practices indicate discursive means wherein software inventions become feminized and eventually widely viable as products whose users themselves become vital commodities. I focus on how Maes grappled with the postfeminist economic and academic structures that helped make her work so influential. At MIT throughout the 1990s, Maes applied artificial intelligence (AI) research to the dilemmas she faced concerning choice, autonomy, and self-management. Born in Belgium in 1961, Maes graduated in 1987 from the Vrije Universiteit Brussel with a PhD in computer science. Her education focused on the intersection of robotics and collaborative learning, the idea that people learn best by working together in a group with diverse perspectives. Following her time as senior research scientist for the Belgian National Science Foundation, in 1989 Rodney Brooks and Marvin Minsky, two preeminent AI researchers, hired her as a visiting professor at MIT in their AI lab. Her research during this period focused on applying collaborative learning techniques to AI, which was also central to Brooks's and Minsky's work.

The great challenge of AI during this period centered on the issue that while one could program a robot with a great deal of knowledge and information on how to complete tasks, if the robot encountered a situation that it was not programmed to handle, it would more than likely break. Maes thus researched collaborative learning techniques that could help AI learn how to do things on its own. Instead of thinking of an AI robot as a single entity, she designed her robots to contain multiple selves that she called *autonomous agents*. Each agent was a small program that kept track of a detail concerning its environment. Together, many different agents interacted within the robot and tried to figure out how to handle different scenarios. For instance, a robot designed to walk might not have been programmed to know how to deal with stairs. However, this robot could be programmed with an agent that kept track of elevation, one that kept track of distance, and one that recorded differences in the force and movement of the robot's leg motors. Through collaborative learning algorithms, Maes helped to design these individual agents to work together to move the robot as a whole in such a way that it could navigate the stairs and other unforeseen hazards.[24]

While working on collaborative filtering techniques, Maes was actively concerned with the changing dynamics for professional women around work and family life. When Maes started at MIT, Brooks was focused on building Genghis, a six-legged robot that he hoped to one day send to Mars. Brooks and Minsky hired Maes specifically to work on this project with the hopes that her collaborative learning techniques would prepare it for the hazards of walking not just on stairs but also on other planets. During this period at MIT (and in academia more generally), women were rarely hired in or encouraged to enter the traditional science, technology, engineering, and mathematics (STEM) fields. A 2002 study on the status of female faculty in the MIT School of Engineering noted that in 1990, two years after Maes was hired, only 5 percent of the faculty were women, and they were marginalized in a myriad of ways, including by receiving fewer grants and conference invitations than their male counterparts. While the percentage of female STEM faculty members doubled from 1990 to 2000 to 10 percent, this consisted of a jump from only 17 out of 357 faculty to 34 out of 348.[25] During this postfeminist moment, the Department of Electrical Engineering and Computer Science, where Maes was originally hired, employed 28 men and no women from 1990 to 1998.

Maes found this environment stifling and became interested in other potential uses for collaborative learning and autonomous agents, especially as they pertained to the emerging space of the internet.[26] In the fall of 1991, she gained a tenure-track position and moved to the MIT Media Laboratory in the School of Architecture. The Media Lab houses a relatively diverse faculty with a wide variety of interdisciplinary applied science research projects that are funded by many large corporations. While not a prerequisite, the research in this lab can be potentially very marketable, and after graduating, student researchers often turn lab projects into businesses. While Maes characterized the male-dominated computer science department as far too critical of unsuccessful experiments and only interested in computer science for the sake of computer science, she described the Media Lab as interdisciplinary, open minded, and primarily concerned with improving people's lives. She also felt that the Media Lab encouraged risk taking and accepted human error and mistakes in ways that those in the AI lab did not.[27]

Maes saw the differences between the AI lab and Media Lab as heavily gendered, and this gendering pushed her toward the Media Lab. As Balsamo has argued, interdisciplinary collaborative environments like the Media Lab, where people of different backgrounds and assumptions come together, encourage participants to take gender, race, and class into account throughout the design process. With inventions ranging from e-readers to MPEG-4 audio players, the Media Lab's entrepreneurial focus on consumer goods, applied sciences, and interdisciplinarity helped lead to the creation and growth of Maes's work on recommendation systems and social networking. Her description of the AI lab and Media Lab largely mirrors how she conceives of the role of gender in academia more generally. When I asked her whether she thought her gender played any role in her research and career, she responded, "I think that on average women approach problems and research in a more pragmatic way, with more common sense. Men are often in love with the technology itself (rather than being excited about how that technology can change people's lives). I think women bring a lot to CS [computer science] research because CS is no longer about 'computing,' it is about social networking, creativity, expression, etc., all areas which women are very good at."[28] Through Maes's association of the AI lab's harsh criticism and focus on technology and theory with the faults of masculinity, the Media Lab becomes synonymous with positive feminine pragmatics, communication, and creativity. For Maes, these descriptions are symptomatic of how gender is enacted and performed within the STEM disciplines of academia. The different labs' projects evince this gender divide, with one focusing on sending a fleet of robots to explore and study other planets and the other focused on consumer goods and services. Historically, consumerism has been the domain of women, and it is not surprising that an academic woman would find a lab working on projects for the consumer marketplace to be welcoming. It is also not surprising that in a postfeminist era that equates female agency with consumerism, the Media Lab would be open to integrating a woman's perspective into its work on consumer technologies.

Despite continued gender divisions within the STEM disciplines, MIT did not pass up the opportunity to display its (post)feminist credentials through Maes's accomplishments. As an attractive young woman with a new family in a consumer-friendly research lab, Maes became a poster "girl" for MIT's newfound appreciation of the "second sex." The products produced by her unit helped maintain and enhance funding for pure science. In many interviews, Maes discussed her experience as a woman at MIT and always stressed that rather than causing her to experience discrimination, her gender had always helped her stand out. She stated, "If you're a woman in a field where there aren't as many women, you get more attention, rather than less. Or more attention than men at the same level. So I haven't experienced it to be a negative thing at all. I have never made a professional distinction between men and women. I've never compared myself with other women. I have always compared myself with everybody else."[29] This logic illustrates the postfeminist contradiction of wanting to see the world as an equal playing field even while being aware that gender does affect one's job prospects, positively or negatively. While Maes attributes her exceptional status to her position as a woman, she had also modeled briefly in college, and her attractiveness almost certainly made her stand out to her male bosses at MIT. However, her status as an object of the male gaze only got her more attention and not necessarily a higher salary or faster career advancement. References in interviews to her status as a wife, a mother, and an ex-model who travels home to Belgium often and does not "live for work" but rather "works to live" tag her with a palatable and quintessentially postfeminist identity that few could adequately inhabit.[30]

This form of femininity was celebrated at MIT in the 1990s. A report titled *A Study on the Status of Women Faculty in Science at MIT* found that while younger female faculty often felt very positive about their roles in the institution, MIT neglected more-senior female faculty in the STEM fields: they were paid less, they received less grant funding for their projects, and they were invited to far fewer conferences than their male counterparts.[31] The same continues to be true of minorities at MIT, who as of 2010 only represented 2.7 percent of the science faculty and were 10 percent less likely to receive tenure than their white counterparts.[32] Through the press and accolades that followed her throughout the 1990s, Maes became a central character in MIT's (as well as academia's more general) push to reimagine itself as an institution hospitable toward (white) women and the new perspectives they could bring to the sciences from which they had for so long been excluded. Maes is a troubling figure who, while clearly extremely qualified for her position, was also treated as a symbol that helped MIT hold on to its elite, or elitist, status.

Maes's subsequent work continued to address the needs of female professionals, drawing on her own experience as a tenured professor with a family. Her move to the Media Lab made it possible for her to work on projects that she felt were not just more connected to her interests but also more concerned with the goals of professional women like herself, both within and outside the academy. Maes founded the MIT Software Agents Working Group with the express purpose of applying her work on robotics to software programs that could help her manage the complexity of her own life within academia.

Her first project in the Media Lab focused on creating a program that could automatically make a work schedule for a user and was specifically designed to help Maes manage her own schedule. The autonomous agents in this program would observe the way Maes scheduled her meetings and keep track of how much time she liked to have between them, when she preferred to have lunch, and whether there were people she always met with at particular times. Eventually, this program was able to schedule her meetings automatically as long as her scheduling habits remained relatively unvaried. This scheduler automated the burdensome processes of self-management and self-discipline that are common for both professional women and men in this postfeminist era. It could learn to schedule work around day care, carpools, and other family obligations. For those like Maes who would find this technology most helpful, this scheduler worked to manage decisions about domesticity and professional life primarily by making choices for the person. While the freedom to self-manage and make one's own choices about work and family appears to be rather simple and straightforward, each of these metachoices is surrounded by a plethora of decisions—both large and small—that must be made in order to make the central choices viable.

By allowing people to avoid having to make these choices concerning their daily schedules, Maes's technology responded to the "problems," "burdens," and "freedoms" of postfeminism, postracialism, and neoliberalism in the lives of men and women of all races. Though feminist discourses concerning the distinctions and connections between work and family life largely focus on how different spaces (the office versus the home) structure this often oppressive and, for women at least, false opposition, postfeminist rhetoric structures this binary around time. While the home has always been a site of domestic labor, digital technologies and many other economic factors have also made the home an important site for telecommuting and other new forms of labor. These popular labor practices belie space-based oppositions between home and work; they instead rely on separations between work and "free" time. In *Life on the Screen*, Turkle specifically uses the example of a new parent when describing how Maes's scheduling software works: "If the agent is in the service of a parent with a young child in day care, it might find itself learning to avoid dinner meetings."[33] For Americans, the need to rethink how to temporally structure one's day around public and private life became much more common in the 1980s and 1990s as more and more women entered the workplace and many men found themselves with more responsibilities in the home.

Notably, only the very affluent can even find a use for Maes's scheduler. A scheduler can only help people arrange their lives if they have some control over when they work, which is not a reality for the vast majority of people. This scheduler only works because the standard nine-to-five work schedule no longer applies to as many people who work either in creative industries or at the bottom rungs of all businesses. As those in the bottom rungs generally have no control over their schedule, those in the new creative industries, like academics, designers, artists, programmers, writers, and inventors, are the ones who would find Maes's scheduler useful. For those in the bottom rungs, who have no control over their work times, such schedulers can have the oppressive effect of being used instead by their bosses to create precarious and uncertain work schedules that can change at a moment's notice. Maes developed this scheduler as a way for affluent individuals to manage the difficulties of these new labor patterns, which are articulated around time rather than discrete gendered spaces that distinguish home and work, private and public.

In the process, affluence and women working became a "problem" for everyone. Precarious or flexible work conditions (like all forms of employment) are highly gendered and are always defined in relation to the family structure. In a 1996 *Wired* article, Negroponte, the founding director (with Jerome B. Wiesner) of the MIT Media Lab, comments on the centrality of gender and class issues to Maes's experience of academia and her research: "In the days before Pattie Maes was a mom, and prior to joining MIT's tenure track, she had plenty of time to browse through stores, newspapers and magazines, even cities, with the hope of discovering some piece of treasure. These days she hasn't the time to explore at a leisurely pace, but even worse, the amount of information and the number of products have expanded almost exponentially. What was merely overload yesterday has become impossible for her today. I must say I have felt this way for years!"[34] MIT used Maes's biography and this narrative of her technological inventions "solving" problems in the workplace for women to extend its brand. Negroponte makes it clear that everyone feels overwhelmed by the modern world, but he specifically points to Maes's position as a mother in academia as what made it ultimately "impossible for her today" to shop and explore, two seemingly equivalent actions here. These pressures are central to academic environments, where one's personal and work time often blur in ways that negatively affect women—who still do the lion's share of the housework—far more than men.[35] Maes expressed this as a personal problem that manifested itself in overfull schedules and the routine forgetting of "names and dates and places and locations of various materials in her overstuffed office."[36] While the scheduler could be used by anyone, it helped professional individuals, and Maes herself, establish a routine and re-create boundaries between public and private life that are disrupted by gendered and racialized neoliberal work conditions.

Negroponte also connects Maes's experience to the dilemma of information overload, in which people become so inundated by choices, possibilities, and information that they are unable to make a decision or holistically make sense of a situation. This experience is common when using search engines, where a simple query can result in millions of web page results. While often associated with the rise of digital information technologies and the internet, information overload is also related to (if not actually caused by) a rise in the number and seeming importance of choices within postfeminist culture. Maes designed her recommendation technologies to decrease information overload and make these choices simpler. They do this by selecting and arranging these choices in a way that makes the hypothetically "best" choice (within a neoliberal culture that highly values free market capitalism) the first one with which a user comes into contact. With this technology, Maes turned these choices into recommendations.

Maes's scheduling software assumes that having more control over personal choices is not a great achievement of feminism or the civil rights movement but rather a burden that can be lifted through technology. However, even as Maes's application lifts this "burden," it also necessitates that users give up basic choices concerning how they organize their day to the autonomous agents that design what they consider to be the best schedule. Here, the autonomous agents convert political issues into technological ones. Such software assumes that neoliberalism and postfeminism are not ideologies but instead common sense. This software is predicated on the belief that the oppressive and anxiety-inducing effects of this worldview can only be managed through technological fixes.

These tensions between the intertwined freedoms and burdens of self-management and choice only increased in Maes's second project, an email-sorting application called Maxims. In a further attempt to help better manage the demands of her professional and family life, Maes worked with her student Max Metral to develop the autonomous agents behind her scheduling program for use with email clients. As Turkle describes it, "An agent keeps track of what mail one throws out (notices of films shown on campus), what mail one looks at first (mail from one's department head and immediate colleagues), and what mail one immediately refiles for future reference (mail about available research grants)."[37] After a training period, during which the agents learn and keep track of a user's habits, Maxims takes over and is able to prioritize a user's mail for him or her. Nearly fifteen years later, these same principles have been adopted by several major email clients, including Google's Gmail priority inbox service.

Maxims took these autonomous agents one step further by having them learn not just from the habits of their specific user but also from the agents of other users. When one user's agent "encountered an email for which it did not have a memory, it would communicate with other Maxims agents in the office, finding out whether a message from, say, Maes' colleague Nicholas Negroponte was given a lot of attention, or just a little."[38] This technique, based on the principles of collaborative learning that Maes studied in graduate school, is called collaborative filtering and makes it possible for agents to know how to deal with previously unexperienced and unforeseen events. For example, a user might receive an email from an autocratic lawyer in Nigeria. This user's Maxims agents would communicate with the agents of other users and see what these other users did with this mail. If other users labeled it as spam, the Maxims agents would communicate this and the new Nigerian email would be sent to the spam folder as well. While often very helpful, this action can be problematic if users at some point change their email-reading habits, wish to start paying more attention to correspondence from people whom they previously were not interested in, or decide to start seeing more films on campus. Thus, Maxims constructs conservative digital communities that preserve the habituation of the status quo.

The Maxims software obscures its role in structuring the limits of user agency by stressing the ways in which users can exercise control over the agents' actions. Users can represent the Maxims agents as cartoon human faces that depict the agent's state of mind. With different emoticon-like facial features, Maxims can indicate that the agent is thinking, is working, does not have enough information, or wants to make a recommendation.[39] These expressive faces show how algorithms work and thereby personalize data mining. If Maxims agrees with the user's actions, it will look pleased, but if it has an alternative recommendation, it will appear pensive. In a project on the effects of representing agents via faces in poker game environments, Maes and Tomoko Koda concluded that users find expressive faces in this application to be "likable, engaging, and comfortable."[40] Maes used these findings to make Maxims' users more accepting of these programs by making the agents' actions more expressive, more affective, and more "human." Maes and Koda felt that an expressive face could show how the algorithms worked so that users could exercise a greater level of control in their use. In the process, they rearticulate collaboration not as an effect of communities but rather as a process of personalization and personification.

Maxims' face also reacts to the user's actions, displaying a smile when it would have made the same decision for the user and a frown when it would have done something differently. This action creates a reversal of roles in the user-agent relationship. While these facial reactions are meant to give the user a sense of how well the agent is at hypothesizing the user's scheduling or email-sorting habits and desires, this feedback also lets users know if the agent disapproves of their scheduling abilities and would do it differently if it could. A preference setting allows users to change how certain an agent has to be before it offers a suggestion. As Turkle explains, "If the agent never offers its own suggestion but always shows a pleased face after the user takes an action, the user can infer that the agent's 'tell me' threshold should be lowered. The agent probably has useful information but is being constrained from sharing it."[41] Like the facial reactions, this ability to change when the agent offers suggestions does not have any effect on the kinds of suggestions the agents would make. However, as an extra level of interpellative self-monitoring for users, these automated reactions do have an effect not just on how they think of these decisions and the process the agent goes through but also on how they as users think of their own actions and decisions.

In addition, to make this system work, collaborative filtering depends on users' exhibiting a high level of consistency in their daily choices. Programs like Maxims cannot handle people who are mercurial in their desires and daily life; they are designed to manage such people into becoming more stable and predictable. This is a prime example of what Lev Manovich considers the interpellative effect of interactive media, in which we are asked "to mistake the structure of somebody else's mind for our own."[42] In this case, the other "mind" is in concert with our own, as while we try to make the emoticon smile by changing our actions, it also tries to change to be more like us. As they become more predictable, the user and the machine meet somewhere in the middle. If a user is completely haphazard in his or her decisions about whether to read and reply to certain emails, Maxims' agents cannot help the user organize his or her email. As a result, these technologies encourage users to adopt unchanging patterns in their everyday lives so as to allow the agents to make these lives easier to manage.

Thus, as a disciplinary tool of interpellation, recommendation technologies encourage users to adopt a lifestyle built on a static sense of self that is practiced through unchanging patterns in their daily lives. This lack of change allows agents to make accurate recommendations and thus authorizes them to manage the lives of users. Hugo Liu, Pattie Maes, and Gloria Davenport link this conception of this static form of subjectivity to the works of Plato and philosopher Georg Simmel in "Unraveling the Taste Fabric of Social Networks."[43]

These theories help Maes also link time-management issues to consumption habits, as they both become expressions of plenitude and lack. In this essay, Liu, Maes, and Davenport focus on Plato's theory that "in a culture of plenitude, a person's identity can only be described as the sum total of what she likes and consumes."[44] McRobbie and other postfeminist scholars echo this sentiment when they discuss how an affluent woman's individuality is constructed out of a consumerist material culture.[45] Yet, while these scholars focus on the plenitude of material goods, the digital plenitude of seemingly endless information, data, and media is just as much a force in this transformation. Furthermore, these various contemporary plenitudes are deeply connected to each other and help to support each other's existence. Many scholars and journalists have commented on how we are now living in an age of abundance in terms of everything from information and consumer goods to food, education, and people. Countering the belief that we are in an era of cultural, environmental, and economic decline, and too often ignoring the tremendous imbalances in access to this abundance, these critics often argue that this abundance is changing the way we think about and perceive the world. Seemingly in response to those like Frank Pasquale who fear how limited our access to information currently is, Mark Andrejevic has argued that in this "infoglut" culture, the problem of information overload assumes that "more information obscures rather than clarifies the picture."[46] A host of self-help books now focuses on managing this sensory overload via techniques like meditation, sensory deprivation tanks, and "information diets." Here, plenitude and information overload is less a natural or innate limit to how much information we can take in, but rather a symptom of our information economy and contemporary culture's habit of framing choice as a burden. In discussing the "lure of a sensory plenitude presumably available simply, instantaneously, and pleasurably with any one of several clicking apparatuses," Anna Everett asks, "Why buy the rhetorics of plenitude or the expensive machines and services of click culture?"[47] These plenitudes situate us as users, laborers, and consumers of digital media and are central to our conception of contemporary culture and how our gendered and racialized bodies and identities are shaped by it.

Maes related Plato's culture of plenitude to her constant grappling with too much information and her effort to organize and help make sense of it for herself and others.[48] The culture of (digital) plenitude affects the categories by which users identify themselves and, as many postmodern theorists have demonstrated, also increases the number of possible identities available to them.[49] Rather than relish this plenitude, Maes views it as a burden. By invoking Plato, she obscures the contemporary and gendered aspects of this problem and portrays it instead as a timeless issue that affects everyone. Her technologies work to simplify and at times obfuscate this plenitude while also presenting it as an essential affliction of humanity rather than one affecting a small subset of professionals in creative industries. This is done specifically by limiting and guiding the user toward this "sum total of what she likes and consumes."[50]

Maes's conception of subjectivity in an era of plenitude and its relationship to contemporary gender expectations is further elucidated by her discussion of Simmel. For Simmel, human identity is like a broken pane of glass: "In each shard, which could be our profession, social status, church membership, or the things we like, we see a partial reflection of our identity. The sum of these shards never fully captures our individuality, but they do begin to approach it."[51] Like Erving Goffman after him, Simmel describes how people present themselves variously in different situations and points out that these different modes of presentation are not false or wrong but are just as important as every other. Maes used Simmel's theories of identity as a foundation for her autonomous agent-based projects, including Maxims, and designed her agents to act as shards that together approach but never completely capture the user's self-image.

In Maes's work using agents to design a user's schedule and filter email, the collection of all of a user's agents becomes a mirror that she argues reveals particular qualities that the user did not realize he or she possessed. Liu, Maes, and Davenport call this mirror effect an "earnest self-actualization," a self-help industry term meaning a tool for the self-management of the user and his or her self-image.[52] She proposes that "actualization" implies that this technology does not just reveal the inner selves of users but in fact helps users to make themselves into the people they want to become. This rhetoric recalls the seemingly endless, nauseating posters, mugs, greeting cards, and other merchandise commanding consumers to "be your best self" or "BYOB: be your best self, only better!" These agents act as a two-way mirror that can be used not just for reflection but also for surveillance and discipline; they operate as tools of self-discipline that shape and reflect a sense of self-managed by the expectations of the contemporary dominant culture. This technology is directly linked to the self-development ideology of neoliberalism and postfeminism that tries to shape subjectivity through the manipulation of individuals' actions and consumption habits.

In these articles, Maes and her co-writers describe their technologies as tools to help users both transform and gain insight into themselves and the cultural materials they were most affected by; however, these agents themselves act as cultural materials that also shape these users specifically as neoliberal consumers. They base their belief that these autonomous agents reflected aspects of the user who made them on the idea that "that common sense tells us that people are not completely arbitrary in what they like or consume, they hold at least partially coherent systems of opinions, personalities, ethics and tastes, so there should be a pattern behind a person's consumerism."[53] Maes's recommendation technologies pay attention to those parts of a person's identity that can be best utilized to make recommendations for everyone. In the process, economic utility becomes synonymous with identity.

Yet Maes believed not that such a pattern of consumerism was always directly mimetic but rather that this parity between one's self and one's possessions is an implicit goal of the postfeminist consumerist society in which she lived and worked. She compares this contemporary subject formation to Aristotle's *"enkrasia* or continence and thus the ability to be consistent."[54] As Timothy O'Connor and Constantine Sandis explain, enkrasia is important because "it is a central requirement of practical rationality" that requires "you to intend to do what you believe you ought to do" in an unconflicted way.[55] Enkrasia is opposed to *akrasia*, or actions that one takes even though one recognizes that they are "in conflict with what she judges to be the best course of action."[56] An akrastic personality is one that is marked by spontaneity and a lack of coherence in the individual's overall actions. In this respect, it is a queer trait that marks a desire to not be defined by a fixed identity. Yet akrasia is often characterized by self-conflict and an inability to make a firm decision or come to a conclusion because no single option appears to be the obvious best, which indicates an underlying desire to be normal, even if that normality is not achieved. These characteristics frame akrasia as a pathology suffered by neoliberal subjects of precariousness for whom no choice is guaranteed to have positive outcomes. From this perspective, akrasia is a problem in consumer environments, where the distinctions between products are often minimal and there is no clear best choice.

In contrast to Donna J. Haraway's vision of cybernetic feminism and her hope that by combining and collaborating with technology we will become more fluid and equal subjects, Maes's self-management technologies work instead to make us more static and exploitable.[57] These algorithms translate our seemingly idiosyncratic, frenetic, and fluid activities, desires, and thoughts into information. Following Claude Shannon, N. Katherine Hayles defines information as a "probability function. . . . It is a pattern, not a presence."[58] Hayles's explanation helps explain how these filters transform us from erratic (or stochastic) subjects into repeatable and therefore predictable patterns.

Maes's self-management technologies supported consumerism by turning akrastic personalities into enkrastic ones by asserting one option as the best. During the later years of neoliberal Reaganomics, enkrasia was discussed as a primal force of consumerism: "Grant McCracken coined the term the Diderot Effect to describe consumers' general compulsions for consistency" and the theory that all of the objects a given consumer owns "should all be of a comparable quality and enjoy a similarly high echelon of taste."[59] In describing a similar effect, Maes uses the example of John, who "buys a new lamp that he really loves more than anything else, but when he places it in his home, he finds that his other possessions are not nearly as dear to him, so he grows unhappy with them and constantly seeks to upgrade all his possessions such that he will no longer cherish one much more than the others."[60] This uniformity gives the user's interests a sense of "aesthetic closure," which is tied to one's sense of self as enkrastic and contained.[61] In turn, this drive for aesthetic closure heightens consumer impulses and encourages people to continually buy more in the hopes that these purchases will, like one more shard of glass, get them closer to this closure. At the same time, the capitalist need to continually produce goods makes this closure always recede into the distance.

Beyond consumerism, this mode of informational profiling has led to great harm, especially for women and minorities. Jasbir K. Puar argues that this contemporary form of profiling, in which "the ocular, affective, and informational are not separate power grids or spheres of control" but rather "work in concert—not synthetically, but as interfacing matrices," connects to a long history of not just gender but also racial profiling.[62] She notes Horace Campbell's argument that the contemporary racial profiling of suspected terrorists in America brings "'yesterday's racial oppression in line with new technologies and the contemporary eugenics movement.'"[63] Building on Campbell's argument, Puar notes that authorities often use information profiles, full of akrastic patterns, to detect not just similarity (as recommendation systems do) but also difference. The New York Police Department, among others, has replaced racial profiling procedures with lists of specific traits, a process that recognizes "that the terrorist (terrorist is brown versus terrorist is unrecognizable) could look like anyone and *do* just like everyone else, but might *seem* something else."[64] In humans and computers alike, "seeming" different is the product of various types and sources of data that together form "interfacing matrices." Puar's description of the alignment of racial and informational profiling illustrates affinities between how recommendation technologies and repressive regimes construct and use profiling data.

After her scheduling and email-sorting applications, Maes's next and most influential project was a music recommendation system built on her collaborative filtering technologies. Released in July 1994 (three years earlier than GroupLens's somewhat similar MovieLens), the first instantiation of this program, named RINGO, was an experiment funded through MIT that worked via email correspondence with users. While it recommended music to users, it did not provide an option to listen to or purchase it. As Maes and Negroponte describe it,

> What RINGO did was simple. It gave you 20-some music titles by name, then asked, one by one, whether you liked it, didn't like it, or knew it at all. That initialized the system with a small DNA of your likes and dislikes. Thereafter, when you asked for a recommendation, the program matched your DNA with that of all the others in the system. When RINGO found the best matches, it searched for music you had not heard, then recommended it. Your musical pleasure was almost certain. And, if none of the

matches were successful, saying so would perfect your string of bits. Next time would be even better.[65]

As with Maxims, here Maes and Negroponte use the metaphor of DNA to collapse consumer taste with the user's sense of self. Through this software, this sense of self is made to appear both natural and static. As with her earlier applications, difficulties in her own everyday life inspired Maes to create RINGO. Dissatisfied by the radio stations in Boston (and the United States more generally), which tended to be "bland, uninspired," and without much diversity, she felt homesick for the eclectic musical offerings from her Brussels home.[66] The perceived accessibility of this music, which was now not tied down to any particular place, allowed Maes to create an emotional connection to a place thousands of miles away. While these technologies appear to make users more static, they make commodities considerably more fluid. Maes, whose parents, siblings, and extended family lived in her birthplace of Belgium, stated that she designed RINGO to help her avoid homesickness and manage her relationship with her distant family. This mobilization of home frames RINGO as both domestic and global—as universally appealing and comforting.

At the same time, RINGO was a precursor of the wide use of collaborative filtering as a challenge to elitist and expert forms of recommendation. Via RINGO, Maes employed the same discourses of antielitist populism with which these algorithms continue to be associated. The World Wide Web is often touted as a tool for making the opinions and knowledge of experts (whether critics, scientists, scholars, etc.) equal with those of amateurs (who may either have less education or simply not be paid for their opinions). While there are certainly huge problems with this leveling in the realm of science (and facts more generally), Chuck Tryon, among others, has argued that when dealing with media desires and opinions, these technologies have arguably led to more diversity and openness in terms of accepted and popular tastes.[67]

Much of this leveling and openness became connected to the World Wide Web's enabling of instant sales and bringing consumerism into the home. This period saw a massive rise in the number of computers sold for in-home use. At the same time, the internet was introduced to the mainstream public, opening the home as an important new consumer space. RINGO exemplified how the gathering of massive amounts of information that computers and the internet made possible could change the consumer marketplace and self-management. As a two-way mirror, RINGO provided useful information on consumer habits that could both encourage other users to purchase more music and be sold to advertisers.

Maes considered the way collaborative filtering worked in RINGO to be a new and automated form of word-of-mouth communication. RINGO allowed people to spread their personal opinions on different music in order to sway the opinions of others. While experts on music could certainly also sign on and state their opinions, RINGO gave an expert's opinion the same weight as that of someone who had not even heard the music on which they were commenting. Maes contrasted this form of communication with that of newspapers, which use a variety of gatekeeping operations to make sure the information they print is "accurate," a standard not usually associated with taste. By modeling her algorithms as a form of word-of-mouth communication, she suggested that such barriers were much less of a factor in the electronic age and that such (ostensibly) disinterested "accuracy" was not an important element in guessing a user's subjective interest in music.[68] In figuring out whether a user might like a product, looking at his or her general history of likes and dislikes and comparing that history to those of other users on the site worked as a much better predictor of future interests than calling on an expert aesthete in the field. These invocations of intimacy, friendliness, and community rather than elitism helped to make these technologies less threatening and encouraged people to adopt them not just in their everyday lives but also in their homes.

This rhetoric covers over the development of highly sophisticated data-mining operations working at the service of the market. By successfully applying word-of-mouth communication models to RINGO, Maes structured the experiences of a general user base around a vernacular discourse in order to make people feel comfortable with revealing information about themselves that could be sold to advertisers or other interested parties. Maes used these data to encourage greater levels of consumerism and create an ad-based online economy. Moreover, RINGO implicitly asserted consumerism as the primary way to structure online communication and build relationships between users.

RINGO was extremely popular, and in March 1995, Maes and several of her students founded Agents, an internet start-up company, in order to market the RINGO software. Through Agents, they also released a new version of RINGO with a web page interface called Firefly.[69] This program attracted the public's interest and introduced collaborative filtering to the structure of the newly emerging commercial World Wide Web. In Firefly, Maes added the ability for users to add new artists and albums to the database. This technique has been used subsequently by a wide variety of websites, ranging from Wikipedia to YouTube. Only a few months after Firefly's inception, the number of albums in the system grew from 575 to 14,500. In 1995 Firefly received the Second Place Prix Ars Electronica Award, specifically because it could help people manage their many media-related choices in a way that "makes full usage of the power of the Web, [on] a system [that] could only exist on the Web."[70]

Functionally, Firefly remained very similar to RINGO except that instead of just music, it also recommended other types of entertainment, including movies. Maes also added personal profiles to Firefly, which allowed people to share information about their location, name, gender, and age with other users.

While these profiles reconstituted a kind of community in an online space, they also made it possible for Agents to gather information about its user base in order to sell that information to marketers. These marketers could then place advertisements on Firefly that were directed at specific users. Nearly a decade later, companies like Google and Facebook would employ these same collaborative filtering tactics and technologies as the profitable backbone of their websites. Firefly ushered in an era in which the burgeoning online industries of choice and self-management began to commodify users by selling information about them to corporations in order to make their products appear more appealing to users at the level of the individual.

In the process, Firefly became one of the first social networking sites by adding the ability for users to chat with others whom Firefly agents thought shared a similar taste in music (and other media). The program employed the information on users' profile pages in order to put users who supplied similar musical and demographic information in touch with each other. Through these profile pages, people could list their favorite books, movies, music, friends, and so on. Users could check out their friends' profiles, find others on the site who shared a similar taste, and chat with strangers in the hopes of creating friendships based on these shared interests and tastes. Thus, Maes designed Firefly not only to help users find new music but also to help them find other people. This dual operation links consumerist desires to the desire for interpersonal relationships via shared information and proposes that a shared interest in particular forms of consumerism should be central to how users create friendships.

This focus on recommendations as the primary forum for self-management and community is central to postfeminist culture and to the social networks and digital technologies that have emerged from it. While many 1970s feminists asserted nonhierarchical collaboration as a feminine form of community and knowledge building, Maes's formulations relied on technology in Firefly to cultivate postfeminist conceptions of these ideals. Maes commented that what she liked the most about collaborative filtering was its potential for "fostering community."[71] She also believed this motivation was distinctly gendered and linked her own desire to work on collaborative filtering to her femininity: "Maybe it's a bit cliché, but I think that women are more interested in building and maintaining communities."[72] Statements like this reveal a telling overlap between the goals of feminism and those of Maes. Yet at the same time her comments essentialize women and reinforce the idea that women are naturally more interested in creating and supporting their community than men are. Through this rhetoric, she made a market for sophisticated technologies and turned them into viable commodities. While social networking is not inherently feminist or feminine, Maes's work shows that it is indeed in dialogue with the complicated changing gender dynamics of our current postfeminist culture.

Moreover, there is also an important difference in how 1970s feminists defined "community" and how Maes conceives of it. This difference helps to articulate the gap between feminism and postfeminism more generally. For earlier feminists, collaboration was a tool of political and cultural consciousness raising and activism with the principle purpose of creating healthy communities of women. In contrast, Firefly creates communities around a shared cultural taste and recommendations, which function within consumerism and capitalist desire. This collapsing of community with consumerism is a common element of how people identify themselves in contemporary American society and has been roundly critiqued by many scholars of postfeminism.[73]

While the Firefly site became popular immediately after going online in 1995, the huge influence of this technology across the internet was cemented by Agents through its secondary business of managing and selling the underlying Firefly technology to other companies. Many of the most influential internet companies of this era, including Barnes and Noble, ZDNet, Launch, and Yahoo, used collaborative filtering technologies and other technologies supplied and managed by Agents. Many other important and popular websites of our current media landscape also incorporated similar forms of collaborative filtering during this period or soon after, including Google, Amazon, Netflix, and Facebook. Maes's product development research showed that employing user data could be very productive and economical for getting users to purchase ever more products and services or to become products themselves in the form of big data.

These data generate information on demographics that companies like Firefly sell to advertisers who have always relied on reception research but now prize it especially highly because of its usefulness in the creation of personalized online spaces. As in the economies of television that came before it, which sold customers as viewers, this selling of customers as information to advertisers creates a situation in which the user base becomes the real product being bought and sold, while the service that the site offers acts as a lure to get customers to supply more personal information. The question of what companies like Firefly do with all of the information on their users is often front-page news as companies and governments try to regulate how and whether personal information can be sold or used in ways other than those for which users expressly gave permission. While Maes designed her technologies to address contemporary postfeminist dilemmas brought on by changing gender roles and expectations by helping affluent people organize their lives, these recommendation systems also frame the identities of users through the economic value of these data. Identity within online consumer spaces has become valued because of (and defined as) this utility.

Maes took social networking a step further in her next project, Yenta, which focused on building an early social networking and matchmaking website for academics. With several students, including her research assistant Leonard Foner, Maes created a program that would recommend users to others based on the same collaborative filtering techniques that were used to sort emails, music, and other media. While Maes and Foner specifically designed Yenta to help academics working on similar research topics easily find each other directly, its name alludes to a long history of matchmaking rituals that all stress the difficulty of finding someone with whom to spend the rest of one's life and the probability that hormone-addled teenagers will make a bad decision. Like earlier yentas, who ostensibly relied on "rational" economic imperatives to create matches, Maes's program matched people based on general interests and similar life plans rather than sudden emotions and crushes in ways that contrast statistical data with affect and experience.[74]

Like Maes's email program, Yenta was created in order to make it easier and less time consuming for creative industry professionals to find a community of like-minded people; however, it did so at the cost of limiting diversity by bringing people into contact only if they were very similar. By using the same algorithms to match people together as she used to match people with consumer goods, Maes illustrates the lure of rational data as a mode of making choices even in matters of desire and love. Importantly, these relationships were created using the same utility model of identity that Maes employed in Firefly and Maxims. Yenta recommended users to others based on how useful they could be to each other, as determined by work interests, in furthering each of their careers. This created partnerships with economic rather than emotional goals. As I will discuss in depth in chapter 4, echoes of this early work by Maes and Foner are visible in many other current dating websites, including eHarmony.

Privacy under Digital Surveillance

In an article for the 150th anniversary special issue of *Scientific American* in 1995, Maes describes the necessity of software agents if "untrained consumers are to employ future computers and networks effectively."[75] Computers during this time only responded to "direct manipulation," wherein nothing happens unless a user gives extremely detailed commands. Maes felt that, like herself, most consumers did not have the time or the inclination to manually input (or learn how to program) all the tasks that they wanted their computers to do for them. This article advertised the pragmatics of Maes's research, digital technologies, and an MIT education.

Through her research, Maes worked to change the dominant paradigm of computing from one of direct manipulation to one of cooperation, wherein "instead of exercising complete control (and taking responsibility for every move the computer makes), people will be engaged in a cooperative process in which both human and computer agents initiate communication, monitor events, and perform tasks to meet a user's goals."[76] Paradoxically, she imagined that these computer agents would simultaneously be thought of as autonomous and as the user's alter egos, or digital proxies. These proxies would all work simultaneously to help navigate users through online spaces, bring information directly to them, buy or sell objects, and even represent users in their absence.[77]

Maes's technologies rely on and enable a neoliberal, postfeminist ideology that celebrates autonomy above all else and collapses identity, agency, and economic utility. Yet at the same time, her focus on these digital proxies as "alter egos" illustrates a fundamental contradiction of these ideologies and how they are currently experienced and enacted: while, as David Harvey has argued, neoliberalism is dependent on the complete autonomy of corporations and capital that often appears to have no limits, this autonomy is always predicated on an ethic of personal responsibility that demands that specific people be held accountable for all mistakes and failures made by this global system.[78] For this technology to work, it must be autonomous, and yet if anything goes wrong, it is unclear whether the programmer, the agent itself, or the agent's owner would be held accountable.

As these agents became important in economic matters, they increasingly generated fears over their security and the privacy of user information. Indeed, Maes's programs and collaborative filtering more generally quickly inspired fears about how digital technologies were destabilizing privacy by seemingly making personal information readily available both to hackers and to corporations. The ways in which the two-way-mirror aspects of these technologies complicate notions of agency and choice have real legal implications. We must put a great deal of trust in such algorithms and their creators, for, as Pasquale points out, we have no real way of knowing what the algorithms are doing, how they work, or what others can do with them.[79] When Maes tested Maxims, users expressed fears concerning whether they could be held responsible for their agents' actions and questioned who actually owns the information that the agents gathered.[80] The term *agent* also brings up the image of the double agent and the possibility that these technologies could be used to spy on others as a form of industrial espionage. In the case of Maxims, one could even program a malicious agent to hide, delete, forward, or change important or confidential emails.

Maes saw these issues surrounding privacy and consumerism in computing as the fundamental philosophical questions of the digital era. To help allay these fears, Agents created the Firefly Passport, a program designed to keep user information, including name, location, gender, user history, and credit card information, online in a safe and secure way. While Firefly's autonomous agents could view and use the information stored in Passport to make recommendations and purchases, each user had control over which other users could see this personal information. While imperfect, this technology acted as a starting place for the securing of personal information on sites where that information was intrinsic to the working of the site itself, such as social networks.

Agents made their Passport technology the standard way by which secure personal information could be transmitted from one website to another. Passport made it possible to link a recommendation site like Firefly to a shopping site like Amazon, where a user could purchase recommended media. Passport also allowed for sites like Firefly to take information from other linked sites to improve recommendations and create more targeted advertising. In order to make these intercorporation transactions work securely, there needed to be a standard way to transmit personal information online. In 1997 Agents, Netscape, and VeriSign worked together to create a set of rules to govern such transmissions, which they called the Open Profiling Standard (OPS). The OPS called for the creation of a standard user profile that would be located on a user's computer. This profile contained information such as the name, age, location, and contact information of the user, and the OPS defined ways in which a user could safely and easily send this profile information to trusted service providers while allowing the user to know exactly how the data would be used. It also offered further protection via the use of "legal and social contracts and agreements" (similar to a modern terms of service agreement) that would stipulate both what information the service provider had to disclose and how this information had to be displayed to the user.[81] In this way, the OPS codified how companies could use Maes's technologies in turning the World Wide Web into a global marketplace.

Rather than being presented as a way to make online consumerism more productive and efficient, the OPS was advertised as the solution to the problem that "as more and more users, and more and more services come to the internet, users are finding themselves increasingly overwhelmed with the richness of new possibilities, and new service providers are discovering that rapidly building systems to fit these users' information needs becomes ever more challenging."[82] Thus, the OPS became an influential document that regulated privacy, surveillance, and the plenitude of choice in such environments. While Firefly and other collaborative filtering programs were originally designed to deal with the glut of information online, their creators inadvertently also had to figure out how to manage all of the people the programs were keeping track of through their product ratings. The selling of this user information without meaningful consent led to many fears concerning user privacy and the need for regulations designed to standardize the marketplace.

As users are turned into consumers and commodities, their identities also become something that can be consumed, bought, sold, and stolen. Privacy fears speak to how digital technologies instrumentalize postfeminism and neoliberalism in order to reduce definitions of the self for economic gain. These definitions serve to industrialize the self as a commodity regulated and standardized through nongovernmental intercorporate privacy agreements like the OPS. In this context, privacy itself becomes a discursive tool that acts to preserve and protect personal value in order to make that value transactional.

While it is necessary to give up a certain amount of personal information about oneself if one wants the "individualized information, entertainment and services" that a site like Firefly can provide, this giving of information also creates what the OPS authors refer to as a "potential threat to individual privacy."[83] At the time of the publication of their public note in 1997 by the World Wide Web Consortium, there were very few "measures or safeguards" in place to offer "an end user any awareness of, or any control over the usage of his or her personal information."[84] Furthermore, the act of constantly filling out personal information on websites in order to get this personalized user experience was tedious and time consuming for both the user and the service provider. The time and money involved in this data collection tended to outweigh any benefits that the process of personalization could offer.[85] Agents helped create the OPS to confront these time constraints and make them profitable within a culture more and more focused on time management and its relationship to self-worth.

While the OPS did not have any legal force behind it and started out as more of a *recommendation* of one particular type of safeguard that could be implemented, it was eventually adopted into the Platform for Privacy Preferences Project (P3P), which attempted to make it easier for websites to declare the intended use of information they collect about browsing users. While the P3P is only one of many early competing protocols designed to protect user information and give users a better sense of how their information gets used, it has been extremely influential in defining what fair practices in online information gathering should look like and what the ethics behind such practices consist of. Even as Mae's technologies helped constrict online choice, these ethics always equate access to more choices, information, and self-monitoring as an inarguable good that must be protected.

In 1998 Microsoft was also working on improving its online consumer experience and took an interest in Firefly Networks' innovations in the standardization of online consumer data and security. It bought the company in April of that year and began to employ many of its core technologies throughout the Microsoft Network. It also almost immediately took the Firefly website offline. While Maes continued teaching at MIT, many others who had been working on the website, including its cofounder, Max Metral, were reassigned to help start work on the Microsoft Passport software, one of the first programs to implement the OPS in a far-reaching way. Now referred to as Windows Live ID, this program allows users to sign on to a variety of websites associated with Microsoft via one profile and one password that are shared by all sites. These same OPS principles also now underlie the software that allows one to use his or her Facebook user ID and password to sign in "securely" to any of a large number of other websites in exchange for Facebook's knowing (and at times revealing) everything the user does on these other sites, including what products the user may have bought and what movies he or she may have seen. The OPS equated online access with surveillance and exchanged the privacy of users for convenience during a period in which the "wasting" of time for "menial" tasks like entering passwords was equated with a waste of the self.

Conclusion: The Afterlife of Firefly

Maes played a major role in developing the recommendations—the lure—that made the utilization and selling of identity in online spaces both tremendously popular and highly profitable. Many major websites now use some form of collaborative filtering and agent technology to help users sort through the data on their sites, and many, like Netflix, continue to offer Firefly-esque systems for recommending media. With the actual Firefly site now long dead and largely forgotten, the Microsoft Live ID is now the most direct remnant of this history. As the web matured and became structured and used more in the way that Maes imagined it would be, Microsoft and many of the other huge technology companies had to deal with the same issues that Maes confronted during her experiments, concerning how to reconceptualize ideas of community, subjectivity, privacy rights, and agency for this digital postfeminist era. Not surprisingly, these companies used Maes's work as a framework for methods of identification for users across the World Wide Web. As a result, much of the rhetoric and many of the original assumptions Maes espoused concerning the dilemmas of this era associated both with digital plenitude and with changing gender roles have become part of the very structure of the World Wide Web and, with it, our daily lives. Maes's work linked gendered and information concerns together in such a way that changes to information policy now affect the way gender is imagined and judged; transforming gender norms affect what we privilege concerning the definitions and uses of information.

Maes received a great deal of press after Firefly was released. This positive reaction directly tied her work to the continued larger effort during the postfeminist era to show women as highly capable of performing in careers that were previously only open to men. While various groups and labs at other academic institutions successfully implemented collaborative filtering and agent technologies around the same time, Maes was continually singled out as unique for her contribution, and these contributions were rhetorically used to promote MIT's new image of gender equality, which they paradoxically connected to exceptionalism. In addition to winning an Ars Electronica prize for Firefly, over the next five years, from 1995 to 2000, she was named a "Global Leader for Tomorrow" by the World Economic Forum; one of the "15 Most Perspicacious

Visionaries" by the Association for Computing Machinery; one of *TIME Digital*'s "Cyber-Elite" and "Top 50 Technological Pioneers of the High-Tech World"; one of *Newsweek*'s "100 Most Important People"; and, obnoxiously, one of *People* magazine's "50 Most Beautiful People," and she received a lifetime achievement award from the Massachusetts Interactive Media Council.[86] While Maes reportedly agreed to be profiled by *People* because she thought it would "be good for readers to see a woman recognized for her brains," this plan backfired. The resulting article largely downplays Maes's scientific credentials, referring to her instead as a "download diva," and misquotes her as saying that MIT "is almost a wasteland in terms of beauty."[87] By opposing attractiveness to intelligence and science, this article enacts what Negra refers to as postfeminist culture's "distorted rendition of choice" between domesticity and work that serves to reinforce "conservative norms as the ultimate best choices in women's lives."[88] Rather than stress the possibility of being both brilliant and attractive, *People* instead argues that women can indeed excel in academia, but only at the cost of giving up their femininity, defined only by their appearance.

By continually framing her projects as efforts to help herself as well as others, female or male, confront many of the new challenges brought on by changing gender roles and related time constraints, Maes shows the deep, but often invisible, connections between postfeminist culture and the digital technologies created within it. As Maes was the most influential proponent of collaborative learning and filtering during the mid-1990s, a formative period for the internet, her focus on solving problems related to labor, self-management, and the burdens of choice is revealing of how the World Wide Web has always been gendered in complex ways. Many factors were involved in making these algorithms a central part of virtually every important and popular consumer-focused website today, from Google to Baidu, as whole global industries and user bases had to accept and adopt them.

Even so, Maes played a central role in shaping and popularizing collaborative filtering through the gender politics of postfeminism and neoliberalism. While her programs were originally designed for working professional women like her, they have become an important tool for shaping all users into consumers whose identities can be bought, sold, and stolen. While her work attempts to address the problems of neoliberalism and postfeminism, it has also ultimately facilitated the growth of these ideologies and an increasingly inequitable digital economy. In the next chapter, I will discuss how users have begun to critique these systems and challenge the ideologies and power structures behind them.

3

Mapping the Stars

• •

TiVo's, Netflix's, and Digg's
Digital Media Distribution and
Talking Back to Algorithms

> Astrology is a language. If you under-
> stand this language, the sky speaks
> to you.
> —Dane Rudhyar

> Astrology is astronomy brought down
> to Earth and applied toward the affairs
> of men.
> —Ralph Waldo Emerson

> Millionaires don't use astrology,
> billionaires do.
> —J. P. Morgan (apocryphal)

Digital recommendation systems and the collaborative filtering algorithms that they are based on are now commonly used across a plethora of industries. Those who construct and work with these and similar predictive algorithms are often called data scientists, a buzzy title that signals a strong desire to legitimize their

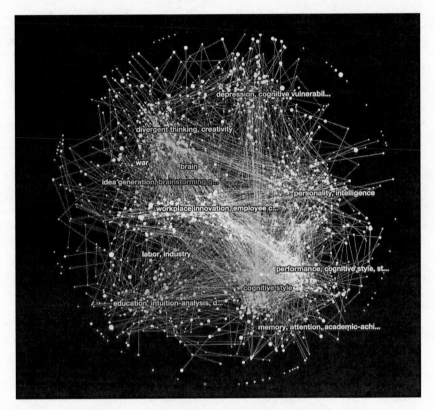

FIG. 3.1 Screenshot of Red Bull's Hacking Creativity project, 2018

findings as objective and unassailable conclusions, extracted (or more often "mined") from mountains of empirical evidence. Data scientists use this mining metaphor to make the information they record appear, like pure diamonds or oil, to have gone through little processing that might throw their validity and worth into question. But like the extraction industries for diamonds and oil, data mining obfuscates all the shaping and reworking that goes into making data scientists' research understandable and relevant.

By attempting to record and analyze all available information, no matter how relevant it appears to be to the problem at hand, the work of data scientists approaches astronomy in its ambition to map and divine the cosmos. This connection is made most evident by the preponderance of data visualizations that illustrate patterns, trends, and correlations in data through the aesthetic of star maps. For instance, Red Bull's Hacking Creativity project purports to "unlock the secrets of the creative process" and find out "what makes us human" based on surveys analyzed by collaborative filters.[1] Red Bull displays its findings in the form of a busy star map that looks to have been attacked by both a paintball gun and corporate jargon like "idea generation," "intuition-analysis," "workplace innovation," and, oddly, "war." This, and the many visualizations like it, connects points of data to each other in constellations meant to accurately, clearly, and undeniably display how everything fits together, what it all means, and where we fit in. Yet, just as one person or culture might look up at the night sky and see a bear where another sees a serving spoon, these visualizations illustrate how subjective and inconclusive the output of these algorithms can be. Rather than astronomy, these efforts are far closer to astrology in their desire to predict our future by mapping out the cosmos.

Collaborative filtering and other recommendation system algorithms are central tools for the analysis of data-mined information. As in other data science applications, recommendation industries argue that with enough information, they will be able to understand us so well that they can accurately know not just what we want but also what we will desire in the future. Yet, just like data visualizations, recommendations obscure far more than they reveal in terms of the complexity of the data they analyze. When we think of these algorithmic recommendations only in terms of whether they get us right, we also risk falling for the same instrumentalist belief that an algorithm (or, for that matter, anything) could ever perfectly capture us and that this is something we should even want. Instead, we should be considering the multitude of ways we interpret and interact with these recommendations, as well as beginning to imagine what creative things we can do with them.

As I argue in my introduction, many scholars and marketers describe algorithms in purely instrumental terms and are concerned only with whether the algorithms work efficiently. Here, I examine how these discourses tend to present algorithms and the recommendations that come from them as deterministically stemming from an objective and natural logic. This logic asserts that these algorithms know us better than we know ourselves and that we should therefore trust their recommendations. However, I focus here on several examples that illustrate that these algorithms are actually quite messy and subjective and can read our data, and in turn be read by us, in a wide variety of ways governed by historical and cultural context. The recommendations these algorithms output may often appear to be stable and clearly understandable, but I use examples from TiVo's, Netflix's, and Digg's recommendation systems in the mid-2000s to illustrate that this is a mirage designed to privilege certain dominant, fetishistic, and deterministic modes of interacting with algorithms. Indeed, we too often ignore or disregard all the weird, wrong, and manipulated recommendations that we are constantly in contact with as aberrant noise. This obfuscation has led not only to our exaggerated fears of and absurd hopes for digital technologies but also to a belief in our own inability to either understand or control them.

Big data industries and those companies that rely on recommendation systems argue that their algorithms can use these mountains of data to reveal objective information about the world and ourselves that was previously invisible to us. For instance, Netflix advertised *House of Cards* as a scientifically guaranteed hit based on its data analytics.[2] Moreover, it gave the impression that the series could not have been envisioned or created without the input of its computer's recommendation; just as Netflix automated the distribution and exhibition of media, it argues that television productions (or at least their concepts) can also be algorithmically generated.

Yet Netflix's story has a number of holes. The most glaring one is that Netflix had little to do with the creation of *House of Cards* and actually bid on it along with premium networks including HBO and Showtime from Media Rights Capital, an independent studio that had previously created comedy series such as *The Ricky Gervais Show*.[3] As Chuck Tryon has noted, purchasing *House of Cards* was less an attempt to bring its audience what they wanted than it was a chance to compete against HBO with prestigious and elitist content of its own.[4] Netflix may have been able to use its data to hypothesize that *House of Cards* would be successful, but this was only after being approached about it and knowing that several other networks were interested. Also, while I do not doubt that Netflix's computers are capable of recommending that it purchase a series like *House of Cards*, is that really that impressive? Hollywood has been using qualitative and quantitative audience data to make production and distribution decisions since at least the 1930s when George Gallup started his famous polls and proposed, even then, that he could eliminate the guesswork of filmmaking.[5] At the same time, are we to assume that humans could not have figured out that since David Fincher, Kevin Spacey, and the original *House of Cards* were all both popular and critically acclaimed, an assemblage of all three might also be successful? Netflix's technology appears to be relying on the same logic of rebooting and mashing up previously successful assets on which Hollywood has always relied. Timothy Havens even argues that, if anything, *House of Cards* is actually evidence for how little Netflix actually knows about its viewers' preferences and how badly its recommendation algorithms actually work.[6] But putting this sick burn aside, we could also hypothetically imagine that Netflix is relying on a very different logic, and the company generated a long list of possible hit ideas; Netflix's executives simply chose to focus on the one that seemed to fit into their rather cliché idea of what a hit could be.

Second, the reporting on this story largely relies on the word of producers and executives at Netflix, who are extremely invested in differentiating their services and algorithmic technologies from both "traditional" Hollywood studios and other digital media distributors. By selling *House of Cards* as a product of their recommendation system, they promote the intelligence of their algorithms and assert that these algorithms (and thereby Netflix more generally) know and respond to their users in order to give those users what they want when they want it.

Last, this rhetoric obfuscates all the other series and films that Netflix produces and distributes that may not present its algorithms in the best light. If its algorithms recommended it make or purchase its less hyped series, such as *Hemlock Grove, Marseille,* or *Flaked,* should we necessarily assume its technologies are intelligent, or just occasionally lucky? Netflix does not release official ratings, and it rarely defines what constitutes a success for the company. In addition to Netflix's assertion that it makes its movies and televisions series based on its algorithmic recommendations, the company also represents its decisions as purely based on objective economic interests. While this rhetoric helps it get away with canceling multicultural ensemble series like *Sense8* and *The Get Down* without many accusations of racism, it does not easily explain why other less discussed series that feature overwhelmingly white casts, like *Flaked, Hemlock Grove,* or *Damnation,* were renewed for more seasons.

My main point here is that the rhetoric around algorithms and recommendation systems is often much more influential than the algorithms themselves. There is a great deal of unexamined hype around the promise of these technologies. But rather than use algorithms to create surprising or bizarre series that no human could have come up with, Netflix and companies like it use their algorithms to uphold a status quo of content largely focused on a white, male demographic and otherwise justify decisions they likely would have made regardless. I hope to begin rethinking how we discuss and use algorithms of all sorts by exploring the underlying assumptions behind how both industries and scholars discuss them while also considering how complicated and messy algorithms and our relationships with them are.

The Myth of Algorithmic Purity

Even algorithmic scholarship is prone to believing the hype around the power and potential of algorithms. Frank Pasquale argues that our world is being unknowingly transformed by the many algorithms all around us, which most of us have no access to.[7] The very foundation of algorithmic culture is premised on the fetishistic belief that what happens in cyberspace and in data is now more important and impactful than what happens when offline. Indeed, as John Cheney-Lippold points out, this logic has thrust us into a world where it is hard to locate areas of culture, business, or politics that are not in some major way shaped by digital technologies.[8] Cheney-Lippold argues that algorithms now often define who we are much more than we do and that this can have very harmful effects. He argues that companies and other entities may get the wrong idea about us from these data and deny our job applications or drop a bomb on us without ever seeing our faces. While I do not disagree that this is possible,

Cheney-Lippold arguably assumes too little of human agency and willpower when he asserts that we are incapable of going around the algorithm or responding to it in ways that its programmers did not perhaps intend. Just because Google thinks I am a Latina or that I like rock concerts does not mean it is right, and it certainly does not mean that those who come across those data (whether human or algorithmic) would necessarily assume that they are accurate. Rather, the fact that these companies and the algorithms they create might just not care about their validity is the problem.

The belief that humans are helpless against algorithms is everywhere. Algorithmic culture scholar Ed Finn and others often present algorithms as extremely powerful and humans as incapable of questioning them. Finn does argue that when algorithms "deploy concepts from the idealized space of computation in messy reality," complex results may occur, but he still preserves algorithms as pure, idyllic, and all-powerful.[9] Algorithms are not apart from the world; they are rather just as complex as (but not more so than) anything else. He also repeatedly compares the algorithm to Neal Stephenson's nam-shub, a Sumerian magical incantation that can rewrite our thoughts and minds, much like a firmware update. Comparing algorithms to magic only makes them appear to be the equivalent of a devilish *suggestion* we are incapable of not following, and in the process, Finn represents humans as extremely *suggestible*.

This rhetoric is premised on the belief that the sheer amount of data that algorithmic output is based on gives it an undeniable connection to objective reality, or what Jacques Lacan referred to as the "real." While each piece of data—whether it be the time one watched a film, the rating one gave it, the number of other films one searched through to find the desired one, and so on—may be meaningless and esoteric, together these pieces of data ostensibly form an indelible bond to the reality they measure. In the parlance of Martin Heidegger or Jacques Derrida, the sheer size of these data sets exhibits a sense of presence of the objects from which they are generated.

In relation to language and writing, Derrida called the belief that the meaning of representations can be completely understood (or determined) first "logocentrism," and later "phallogocentrism." In *Of Grammatology*, Derrida defines *logocentrism* as the premise that speech is more central to language than writing due to its connection to the presence, thoughts, and intent of the speaker.[10] Speech appears to be less or even unmediated in comparison to writing, which logocentrism frames as secondary and derivative of speech (which is itself secondary to thoughts). Yet, in this privileging, Derrida argues, logocentrism illustrates a desire for a form of communication that can perfectly express the thoughts of the speaker to listeners, who will in turn understand them clearly and completely.

Logocentrism is premised on the existence of Logos, the word of God, or divine reason, from which everything stems. In her introduction to *Of*

Grammatology, Gayatri Chakravorty Spivak argues this logocentrism rests on the belief that "the first and last things are the Logos, the Word, the Divine Mind, the infinite understanding of God, an infinitely creative subjectivity, and, closer to our time, the self-presence of full self-consciousness."[11] With this reason comes a belief in an ultimate meaning and order to our lives and the universe. In practice, "the logocentric longing par excellence is to distinguish one thing from another";[12] it creates a desire for rigorous order and stability that strict classifications, definitions, and oppositions appear to provide. Throughout his oeuvre, Derrida considers these binaries, asking whether speech/writing, capitalist/communist, mind/body, self/other, female/male, human/animal, and white/black are far more unstable than logocentrism allows for. In *Limited Inc*, he argues, "All metaphysicians, from Plato to Rousseau, Descartes to Husserl, have proceeded in this way, conceiving good to be before evil, the positive before the negative, the pure before the impure, the simple before the complex, the essential before the accidental, the imitated before the imitation, etc. And this is not just one metaphysical gesture among others, it is the metaphysical exigency, that which has been the most constant, most profound and most potent."[13] Simultaneously, he argues that the creation of these binaries and referral back to divine reason continually privilege certain—largely "masculine"—identities and values over those typically viewed as feminine: the creation and celebration of these binaries is an oppressive and misogynistic force and therefore not just logocentric but rather phallogocentric (a portmanteau of the portmanteaus *phallocentric* and *logocentric* that first appears in "Plato's Pharmacy").

While these properties have long been discussed in relation to language, there is some question over whether they apply to algorithms and their output. N. Katherine Hayles has argued that phallogocentric binaries no longer dominate due to algorithmic technologies and culture. She argues that, rather than resting on the hierarchical binary of presence/absence, digital culture is premised on the new hierarchy of pattern/randomness. The assumption of a Logos that could "give order and meaning to the trajectory of history" and therefore allow for a "stable, coherent self that could witness and testify to a stable, coherent reality" undergirds the presence/absence dialectic.[14] In contrast, she argues the pattern/randomness dialectic does not assume a Logos-like origin and does not attempt to ground signification: "Rather than proceeding along a trajectory toward a known end, such systems evolve toward an open future marked by contingency and unpredictability. Meaning is not guaranteed by a coherent origin; rather, it is made possible (but not inevitable) by the blind force of evolution finding workable solutions within given parameters."[15] Hayles believes (or perhaps simply hopes) that this transformation signals the "end of a certain conception of the human, a conception that may have applied, at best, to that fraction of humanity who had the wealth, power, and leisure to conceptualize themselves as autonomous beings exercising their will through individual agency and choice."[16]

This belief that digital technologies will transform what it means to be human has not, as of yet, been borne out. As I discussed throughout the last chapters, questions of human agency, ownership, choice, and autonomy have remained central to the development of digital technologies. While Hayles argues that autonomous agents and algorithms more generally break down our conception of a "coherent" self, Cheney-Lippold's work illustrates how we continue to invest our data and the patterns that emerge from them with a sense of not just coherence but also ownership, agency, and presence.[17] Like Hayles, he argues that we are constantly made aware of how our algorithmic selves do not necessarily cohere to our conception of ourselves. Cheney-Lippold argues that as we begin to become aware of how our algorithmic selves are different from how we think of ourselves, we also begin to "critically understand the logic by which we are algorithmically valued," which can allow for real resistance to the privileging of algorithmic logic.[18]

I would argue that rather than escaping phallogocentrism, pattern/randomness has become itself a phallogocentric construct. Patterns are defined as containing some sort of meaning, whereas randomness, or noise, is meaningless, or absence—the same binary that structures phallogocentrism structures our algorithmic logic. Thus, pattern/randomness does not contradict but rather maps well onto the presence/absence binary and, if anything, illustrates how pernicious phallogocentrism is within digital culture. Indeed, the pattern/randomness dialectic is also still often fraught with the phallogocentric tendency that Heidegger argued privileges the simple presence of the pattern rather than analyzes the conditions for its appearance.[19] This tendency is most evident in the continued insistence that artificial intelligence machine-learning algorithms can answer questions, find patterns, and make discoveries in ways that *Wired* reporter David Weinberger argues are "beyond human comprehension."[20] The pattern's presence exceeds our understanding of both its meaning and how it was generated. This belief in artificial intelligence's supremacy rests on a phallogocentric—and authoritarian—logic that argues that resistance is futile. While Weinberger (and Hayles) argues that these systems illustrate both that we no longer need to know "how the world works" and that knowing is no longer even possible, it seems a bit premature to argue that we should give up trying. The world has always been too large to grasp, but sketching out the underlying structural logics at play has also been a powerful tool for locating and combating inequities. When we take an algorithm's view of the world at face value, we give up far too much of our ability to critique the algorithm's assumptions and fight whatever inequities it might be blind to.

Algorithms may be different from language, but we still appear to conceive of and shape them from a phallogocentric perspective, or one that believes that the purpose of algorithms is to create perfect communication and express an inarguable or divine order (Logos). Indeed, many philosophers and critics from Maurice Merleau-Ponty to Michel Serres argue that phallogocentrism already treats words as if they are algorithms.[21] Finn directly compares algorithms to Logos in that they both "do not merely describe the world but make it."[22] He then uses this comparison as a structuring device throughout his work to explain the effect we imagine algorithms have on us. In the process, this comparison uncritically portrays algorithms as deterministic and frames the math that undergirds them as transcendental. Algorithms are not ethereal; they spring forth from humans and human-created technologies and they stay put within our earthly realm. They are not a pure form of communication, and they are as open to interpretation, critique, and resistance as anything else.

Algorithms, Identity, and Questions of Privacy

The desire for algorithms to output perfectly transparent information has been foundational to how we imagine our relationship to digital culture and media. Many of the debates that have carried over from the beginnings of the internet concern how and whether we are "present" within these technologies and how we can best communicate our ideas through them. While phallogocentrism connects the presence of speakers to their spoken words by focusing on how their intonation and gestures help shape and ground the speaker's meaning (in ways that written words do not), in the context of algorithms, this logic works to ensure a sense of presence and connection through the recording of huge amounts of metadata (data about data, or the "gestures" of data). For example, the reader of an email (whether a person or an algorithm) may not be able to tell whether it is meant to be sarcastic, but if the reader also has access to the time the email was sent, the place it was sent from, the list of people it was sent to, what device it was written on, and so on, as well as similar information concerning every other email the author has ever sent, the reader may conceivably have a better understanding of not just what the email says but also what the author means.

Yet, even in the 1980s, many feared that metadata and the sense of presence that it affords could reveal too much about individuals and violate their right to privacy. This question of whether metadata has an objective and inherent meaning that can reveal something hidden within us lies at the heart of the Video Privacy Protection Act (VPPA), one of the United States' earliest attempts to legislate big data industries. During the ill-fated United States Supreme Court confirmation hearing of Robert Bork, Michael Dolan at the *Washington City Paper* discovered where Bork rented videos and got a copy of his rental history (an analog form of metadata) from the clerk with the understanding that "the only way to figure out what someone is like is to examine what that someone likes."[23] He further wondered what it would mean if "Robert Bork only rented homosexual porn ... or slasher flicks ... or (the ... horror ...) Disney."[24] In fact, Dolan ended up finding a banal rental history that indicated only that Bork was something of a cinephile, having rented 146 films in two years, and that he particularly enjoyed films directed by Alfred Hitchcock or that starred Cary Grant.[25] Indeed, it seemed that his metadata did not shed any light on Bork whatsoever. The rental history could indicate anything from a simple interest in snappy dialogue to (as Tania Modleski noted shortly after this event) an identification with threatened women and a concomitant desire to punish them.[26] Even though these video store records ultimately appeared not to act as proof of anything concerning Bork's inner life, the tag line of Dolan's article salaciously suggested they did: "Never mind his writings on Roe v. Wade. The inner workings of Robert Bork's mind are revealed by the videos he rents."[27] The story gained national attention due primarily to its reliance on the video rental history—a type of data that many assumed was private. Indeed, many believe that, at the very least, this story added fuel to the fire, making it more difficult for Bork to be voted into office as the privacy of his video rental histories became linked to the much broader fight over the privacy arguments of *Roe v. Wade* and a woman's right to choose.

As Tryon has noted, Dolan's story sparked a great deal of fear in a Congress whose members realized they might also have their video rental histories (presumably chock full of sketchy pornography) broadcast to the general public.[28] Congress almost immediately passed the federal VPPA in 1988, which made the disclosure of rental information to the public, markets, and the police a crime, punishable by not less than a $2,500 fine per video unless the consumer had specifically consented or warrants had been issued.[29] This act also required video stores to destroy rental records no later than a year after the termination of an account.

The VPPA was voted on just as video stores were beginning to store their customer data not just for their financial records but also for potential future monetization. While Vans Stevenson, on behalf of the Video Software Dealers Association, a national organization of video distributors and retailers, praised the law for clarifying the rights and responsibilities of video store clerks, especially in relation to criminal proceedings, Richard Barton, on behalf of the Direct Marketing Association, feared it would hamper their agendas.[30] Just as Web 2.0 companies now collect data on the buying habits of users in order to advertise to them more effectively, marketing firms in the late 1980s were collecting and selling similar data to create mailing lists and other personalized advertisements. This industry, which had always been charged with regulating itself, sought exemptions that would allow video stores to sell information about their customers' habits. The marketing industry also wanted the privacy of these data to be an opt-in rather than opt-out policy, which would assume customers did not care about their privacy unless they expressly stated otherwise.[31]

Senator Patrick Leahy responded that practices like direct marketing via mail lists were exactly the kinds of practices that this law was trying to rein in, and marketers were a primary group that should not have access to such data.[32] As an example, he discussed his young teenage son, who sent away for a "manual on karate or kung fu or whatever all the kids were doing at the time" and was added to mailing lists without his consent.[33] As a result, he started receiving mail ads that "ranged from Soldier of Fortune type ads to some of the most prurient lingerie things I have ever seen. There were ads for X-rated books, videos, and something that came very close to how to kill your neighbor. And I'm serious, it seems to have triggered all of these things."[34] He feared that these practices would only become more sophisticated "with computers and with profiles of people. And politicians do this, too, on profiles of voters and everything else. But how much should we be allowed to profile somebody?"[35] Leahy echoed the fear that direct marketing, a precursor to modern big data industries, would turn his son into a pervy serial killer and also seemed entirely aware of how silly that possibility was. Yet, prescient of things to come, Leahy pondered,

> What happens with interactive television when, in the next generation we are going to be doing so much more by using telephone lines, and televisions to pay bills and buy foods, and maybe run the lights in our houses. If somebody wants to spin out the Orwellian theory, you can have this view of knowing what time I leave the house, what time I come back, how much I get paid, when I get paid, whether I pay my bills on time or I am late on some others, what I like to eat, what I am entertaining and everything. You know, it is almost like having somebody in the dark with binoculars sitting outside your house and it is a little bit chilling.[36]

Barton responded that, while such practices may make some people uncomfortable, no privacy had been invaded by these acts because no one person in any of these marketing companies would have known that Leahy's child had received pornographic literature. These choices are all made automatically via computers, and the data are stored on magnetic reels that are difficult, if not impossible, for actual people to search through.

This debate has not ceased, and we continue to question the value of privacy and weigh the trade-offs of technological conveniences. Recommendations as both a practice and a technology continue to be at the center of this debate. The idea that automated uses of personal information are not a privacy breach is particularly prevalent today: companies constantly store, use, and sell our private data, but—since users assume that no people ever actually see the data—

neither these users nor the companies often recognize these transactions as a breach. These assumptions protect companies, which can always blame privacy breaches on technical snafus or a lone employee instead of the very structure of their industry. Yet, given the invasive actions that marketing and Web 2.0 companies accomplish as a matter of course, the image of "somebody in the dark with binoculars sitting outside your house" makes these companies appear almost respectful in comparison.

Leahy explained the necessity of the VPPA for a rapidly approaching "era of interactive television cables, the growth of computer checking and check-out counters, of security systems and telephones, all lodged together in computers."[37] This act became the first to attempt to protect the privacy of personal data in the face of new media and digital environments that could make all such data public. The act quickly made video rental records one of the only protected forms of consumer information.[38] It even provides more protections to video rental privacy than the 1996 U.S. Health Insurance Privacy and Portability Act affords to health records, as it allows consumers to sue offenders directly. While this belief that video rental records may reveal our true selves illustrates how powerful and common phallogocentric fears can be, it also shows that they are never as stable as they may appear. Leahy is wary of the future power of computers, but as of yet, the only examples he can point to show how dull and random metadata typically is.

At the same time, what data are considered a pattern and what data remain random noise is culturally and historically defined and can change quite quickly. For example, during our supposedly postfeminist and postracial moment, we are continually tracing data points in an attempt to answer questions such as, What does race (or gender, or sexuality, or ethnicity, etc.) have to do with it? What number of police shootings of black men constitutes a "pattern" of racism? How many surveys need to be taken before we can say whether sexism decided the 2016 presidential election? How long does it take to decide whether "it gets better"? Whether one identifies a pattern or just random noise is dependent on a host of culturally dependent conditions and contexts that are arguably more important to consider than the pattern itself.

Like Finn, I believe it is important to develop "algorithmic reading" strategies (a phrase that implicitly links algorithms to language) and caution against taking at face value how algorithms are marketed or the ostensibly dominant meaning of their output.[39] Instead, I would argue for the development of critical analysis that highlights the complexities and incongruities present at every step of an algorithmic apparatus, from when humans originally record and enter data, to when algorithms process them, to when humans later read the final output. Too often, humans are distanced or forgotten in this process, which makes the algorithms they create appear far more like black boxes than they actually are. At each stage, humans—and therefore the programs they write—make "errors," misreadings, mistranslations, and subjective or random decisions; at each stage, historical context and cultural assumptions can shape how and what algorithms calculate. While these moments are typically considered "errors" that will, as technology progresses, be stamped out, I would argue that these subjective moments are critical and undervalued parts of algorithms that are worth exploring for what they illustrate about the messiness of algorithmic culture itself. Algorithmic technologies are not pure or perfect, and the communications that come from them are far from Logos; instead, they are profoundly unmagical and mundane. For the rest of this chapter, I will discuss examples that illustrate the importance of foregrounding these undervalued aspects of algorithms and the culture they influence.

Algorithmic Anxieties (My TiVo Thinks I'm Gay)

What happens when your algorithmic recommendations think you are gay when you are not?[40] Are we, as Cheney-Lippold argues, "pushed and pulled by different interpretations of our data through different algorithmic instructions," or can we push back against the algorithms themselves?[41] In 2002, during Silicon Valley's recovery after the dot-com crash, at the same time that a push for sexual equality took off in the United States and across the globe, a number of newspaper articles, blog posts, comment boards, television episodes, T-shirts, and other media all at once began pondering the question of what to do if TiVo "thinks you are gay." This event helps illuminate how complex this push and pull of identity can become in relation to recommendation systems.

TiVo, the first personal video recorder (PVR), released in 1999, had become famous for allowing users to easily record any content on their televisions and fast-forward through commercials. In addition, TiVo's PVRs kept track of what users watched and used this information to automatically record similar content for users' enjoyment. With this surveillant tracking, TiVo popularized the use of the "digital recommendation system," a collection of algorithms that automatically propose to users various types of media based on other content that they have enjoyed in the past. These systems proved both useful and appealing. Yet their recommendations also generated and revealed certain cultural anxieties having to do with the relations between taste, consumption, and sexual identity.

Digital recommendation systems guess at and, in turn, work to reveal who we are as digital consumers and subjects. This practice can often lead to an algorithmic form of "outing," as "Weird Al" Yankovic complained:

> But I only watched *Will and Grace* one time one day
> Wish I hadn't 'cause TiVo now thinks I'm gay.[42]

With these automated recommendations, TiVo was at the forefront of the personalization industry and its attempts to automatically customize media experiences (whether on television, on the internet, or via other routes) for individual consumers. Although Yankovic's song indicates that his recommendations are a result of his (regretted) viewing habits, Lisa Parks inverts this relationship and refers to such programming practices as "the programming of the self."[43] She argues that though personalization technologies are heralded for expanding viewer choice by introducing us to content we may not have been familiar with, these choices are always "clearly circumscribed by marketers' determinations of 'relevant' content" and thus have "less to do with the viewer's personhood and more to do with new industrial structures of individuation geared toward profit making."[44] TiVo does not simply mirror the personalities of users back to them; it does not just recommend content but, more importantly, implicitly imposes potential identities. Indeed, TiVo's slogan, "TiVo Gets Me," connotes not only its deep—and surveillant—understanding of customers but also its potential control of these customers, as in "TiVo gets me to . . ."

As a digital technology designed to help users bypass commercials, TiVo appeared to disrupt the status quo of the television industry. Yet it was also deeply enmeshed in this industry and its larger efforts to remain dominant within a digital landscape. Indeed, various companies including AOL, Time Warner, Microsoft, Apple, and WebTV had been developing "interactive televisions" (iTVs) throughout the 1990s at tremendous costs. While, as William Boddy has noted, many of these failed efforts focused on making the television more like a simple computer for reading email and surfing the web, TiVo succeeded by presenting itself not as a computer but rather as an updated television.[45] With its simple interface and status as a new, digital version of a videocassette recorder, it primarily became a tool for making television viewing easier and more manageable.

Indeed, though television viewing tends to be framed as passive, the large increase in television channels (from, on average, 41.1 receivable channels in 1995 to 102.1 in 2002) and content from the late 1990s through today had made the act of picking something to watch an often complicated decision that TiVo proposed to ameliorate.[46] Rather than depicting television as a passive (dumb) medium, TiVo presented it as a complicated landscape that even the savviest viewer could not easily navigate. One TiVo commercial from 2000 uses a first-person perspective to display the viewer calmly walking through a television network office with two bodyguards, who (equally calmly) toss an executive out a window; the commercial then commands the viewer, "Program your own TV network."[47] With the ever-increasing amount of content on television, TiVo asserts in the commercial that we—regardless of our sexuality, gender, race, age, or class—must demand that this content be both more heterogeneous and more personalized. As "you," in first-person perspective, are barely noticed by any of the workers in the office when approaching the executive, this commercial simultaneously proposes that television networks do not respect viewers and that this lack of regard has specifically led to the need for viewers to actively take control of their media—to overthrow the system.

These efforts to present itself as subversive and revolutionary worked, and TiVo quickly became synonymous with interactive television: for a while, the term TiVo even became both a general term for all PVRs and a verb that referred to the act of recording something on a PVR.[48] At the same time, TiVo presented itself as the consummation of a long trend in the television industry from broadcasting to narrowcasting to microcasting. Ron Becker has argued that since the 1970s, television marketing has "broken Americans up into narrower and narrower psychographic groups targeted through direct mail, lifestyle magazines, and cable channels."[49] As American culture became more fragmented, advertisers began targeting their messages to niche audiences they believed would most likely be interested in their products. Recommendation systems are, of course, a central technology of this trend.

As television marketers fixated ever more narrowly on their search for lucrative eyeballs, they were guided not just by class, race, and gender divisions but also by other splits in contemporary culture: most notably, those related to sexuality. Becker argues that sexual identity was central to a transformation in advertising and that niche marketing led to a huge increase in homosexual narratives on television. His work, which largely focuses on the period right before TiVo's PVR, helps to situate TiVo in relation to the television industry and the connections it made between narrowcasting and the greater visibility of gays on the networks. In the 1990s marketers and television networks focused on the "slumpy" demographic (socially liberal, urban-minded professionals) as the most valuable niche audience.[50] Networks tried attracting this group by airing gay characters and plotlines designed to appeal to their liberal, multicultural, and "hip" politics; by 1997 there were thirty-three queer characters on prime-time network television.[51] Notably, much of this programming was directed not at a gay audience but rather at the gay-friendly, but straight, audience "looking to reconcile their liberal ideals with their bourgeois materialism."[52] Gay content, which in the 1990s was rarely ever actually sexual, was sold as edgy, but never "over the line."

At the same time, niche marketers were also becoming more interested in gay and lesbian demographics because of their perceived high value as consumers—a perception that, Becker points out, was exacerbated by highly questionable research methods and grossly distorted statistics.[53] In the process, they helped shape which homosexual and queer identities could be made legible. Like Becker, Lisa Henderson describes in detail the complicated ways in which class structures act as a dominant force in how gay—often specifically queer—communities form and how homosexuality is understood. Class distinctions shape queer identity and define what characteristics are positive (wealth, family, normalcy, and "fabulousness") and which are negative ("performative excess and failures of physical control") in ways that "appeal to a range of consumers and still flatter those at the crest of advertising trade value."[54] In this way, homosexuality became a legitimate and legible identity in mainstream culture because of its connection to upward class mobility and consumerism—a transformation that actually ignored the vast majority of gay people.

Although many narratives used glamorized images of what Henderson calls "good queerness" to reinstantiate the dominance of heteronormative values, the "my TiVo thinks I'm gay" plotline illustrates how they also simultaneously depicted a breakdown between heterosexual and homosexual identities and desires.[55] "My TiVo thinks I'm gay" relied on the "mistaken sexual identity plot twist"—one of the most popular gay narrative devices of the 1990s—in which a gay character is mistaken for being straight, or vice versa. As the question in this plot was always, What should you do if TiVo thinks you are gay? it is clear that though these plotlines may have involved gay content, they were always aimed at a straight audience anxious about its own sexuality and privilege. As microcasting became ever more targeted toward specific people rather than groups or demographics, questions of identity construction became even more relevant.

Becker posits that these narratives reinforced "the notion that the line between gay and straight was far from clear."[56] Although they taught the lesson that we all share the same basic humanity, regardless of sexuality, they did so by marginalizing any queer politics or desires to claim any differences between gay and straight cultures. These narratives and the increasing normalization of gay identities created what Becker calls a "straight panic [that] refers to the growing anxiety of a heterosexual culture and straight individuals confronting the shifting social landscape where categories of sexual identity were repeatedly scrutinized and traditional moral hierarchies regulating sexuality were challenged."[57] As these narratives began to not just validate but also valorize homosexual identities throughout the late 1990s, they destabilized the privileged normative position of heteromasculinity.

Fictional accounts of TiVo users focus on themes of straight panic and sexual identity confusion, reflecting the ways in which TiVo and many other algorithmic technologies present themselves as transgressive while simultaneously reinforcing certain dominant, heterosexual norms. For instance, in an episode of the HBO series *The Mind of the Married Man* directed by Bruce Paltrow, TiVo is playfully framed as a menace to patriarchal, retrograde definitions of heteronormativity—a position that HBO itself and much "quality" television have at times and with limited success tried to occupy throughout the history of narrow- and microcasting with series like *Sex and the City* (1998–2004), *Six Feet Under* (2001–2005), *The Wire* (2002–2008), *The L Word* (2004–2009), and *Looking* (2014–2015).[58] This *Mind of the Married Man* episode features a TiVo introduced by its ability to know that "if you record *Star Trek*, [it] assumes you like that kind of thing and then when you aren't home it records *The X-Files*." This product placement is highly self-conscious both because TiVo symbolized a type of luxurious spectatorship that many series wanted to be associated with and because TiVo's fast-forward feature was itself a major cause of the growth of product placement on television more generally.

At the beginning of the episode, the protagonist, Micky Barnes (Mike Binder), announces to his coworkers that because he recorded a few episodes of *Will and Grace* and *Ellen*, TiVo now thinks he is gay and will not stop recording *Queer as Folk* and Judy Garland specials. Micky explains that he has been trying to outfox the program to "get it to go the other way" by recording *MTV Spring Break* (2001), *Playboy after Dark* (1969–1970), and *MTV Nipple Parade* (a series I had to look up to make sure it was not real). Unfortunately for Micky, "the thing won't budge." He complains, "It insists I'm gay; it's a problem." TiVo's recommendations become a point of crisis for Micky as he goes through the rest of the episode considering what the relationship between his actions and his perception of himself really is. Throughout the episode, Micky's TiVo recommendations force him to question himself and his relationships with his friends and colleagues. Here, Micky illustrates how "straight panic" became attached to TiVo and microcasting strategies more generally. This was a strategic effort to get the slumpy demographic not just to use PVRs but also to care what their technology "thinks" of them. As this metanarrative focuses entirely on television viewing habits, it also emphasizes the changing media landscape and vexed reaction by HBO and other premium outlets that quite suddenly catered to many new and previously marginalized audiences, including the LGTBQ community, while featuring series driven by testosterone (and full-frontal nudity) like *Hung* (2009–2011) and *Entourage* (2004–2011).

This issue recurs in an episode of *King of Queens* wherein TiVo's efforts at recommending media are characterized as a form of sexual harassment. This episode, called "Mammary Lane," presents three farcical sexual harassment scenarios that together lampoon victims of real harassment.[59] These narratives include those of a toddler who continually grabs her babysitter's breasts, a man who tries to be nice to an unattractive bowling alley attendant by attempting to flirt with her, and a TiVo that records gay-themed series for its heterosexual owner and thus ruins his date. The victims of such "harassment" are shown in different states of trauma, with one fearing for her life in a parking lot and another rocking back and forth on her bed while she is accused of "asking for it" from a toddler. On the one hand, these three narratives depict a postfeminist backlash or hatred of feminism (if not women more generally) that indicates how TiVo's suggestions can harass its users, while also making fun of the serious import of harassment. Imputing harassment to a machine and a baby underscores ways in which harassment can be framed as inadvertent and as a misinterpretation of meaning; the harassers here are innocent, while the harassed are crazy. On the other hand, these scenarios are so outrageously silly and self-conscious that they taunt the viewer to analyze the situation's comedy and critique the postfeminist rhetoric of harassment and victimization on which it relies. In either case, the episode presents as absurd the fear that TiVo's surveillance strategies may have unforeseen consequences and that its "opinion" of users would be trusted more than the users themselves. Rather than phallogocentric, moments such as this illustrate algorithms as simply another perspective that is just as prone to error and misreading as any other.

The episode introduces TiVo through Spence Olchin (Patton Oswalt), who has just purchased it for his bachelor pad. In an attempt to impress a female neighbor he has a crush on, Spence asks her, "Want to see my new purchase? It's my TiVo; it's this thing that records TV shows but the cool part is that it is intuitive. Once you program in some shows you like, it gets to know your taste and then, automatically picks out other shows for you." After explaining that he programmed in "*Sex and the City*, *Six Feet Under* and a few other favorites," he checks to see what TiVo recorded for him automatically and finds *The Adventures of Priscilla, Queen of the Desert* (1994); *Judy Garland, Live at Carnegie Hall* (1961); *Decorating with Style* (1996–2002); and *Queer as Folk* (2000–2005). In a moment of (straight) panic, Spence yells, "Oh my God, TiVo thinks I'm gay!" His attractive neighbor then abandons him to his heteroanxiety. TiVo appears to have an agency of its own, which makes Spence uncertain as to whether the machine is judging him or whether it is instead simply mimetically reacting to his own actions.

In an odd turn, TiVo becomes a status symbol even as it makes Spence extremely uncomfortable. Along with associating homosexuality with upward mobility, the implication that homosexuals use TiVo makes the technology—and contemporary television more generally—appear classy. Spence makes his neighbor check out his suggestions because he hopes that she will be impressed both by his expensive new machine and by the suggestions themselves, which he hopes will portray him as the interesting heterosexual that he imagines himself to be. Although Spence challenges the suggestions TiVo has for him, he never questions TiVo's role as a self-representational technology, even though he does not like the representation it has come up with. Yet he questions the generated self-representation by imparting TiVo with agency. By placing these anxieties and the threat of TiVo suggestions to normative conceptions of sexual identity within such a benign context, they are both made to appear hyperbolic and silly. This situation nods to the reality that TiVo can indeed create anxiety about a user's sense of sexual identity, while also making clear that this is not an issue that should be taken seriously. The humor of this episode leaves the troubling nature of this recommendation system intact.

Along with sexual anxiety, TiVo's appearance here also points to class pressures that are an explicit part of *King of Queens* as a whole. Henderson outlines how, in both broadcast and niche television, representations of homosexuality tend to "borrow liberally from the class fantasies of everyday life, especially fantasies of mobility and having."[60] This process is inverted on *King of Queens*, as Spence is classed in part due to his queered desires. The characters in the series are largely high school–educated, white, heterosexual members of the lower to middle class with aspirations for more. Spence himself is a nerdy tollbooth worker from rural West Virginia. With an interest in science fiction and comic books, he is perhaps the most cultured of the series' ensemble. These interests are underscored by his declared penchant for HBO "quality" series, which Michael Z. Newman and Elana Levine argue constitutes a complete rejection of the style, address, and viewer positioning of *King of Queens* and other similar multicam sitcoms associated with a dated broadcasting strategy.[61] They also point out that the television industry used PVRs to legitimate itself as a higher art form by suggesting that it was no longer just for the "'saps' and 'dipshits' who tune in each week to follow a narrative, with nothing better to do than to be at home, in front of the set, sitting through the commercials."[62] They demonstrate how narrowcasting, microcasting, and video recording have always been classed. Yet, with its prodigious use of product placement, coupled with an anxious depiction of PVRs, *King of Queens* illustrates the vexed nature of these legitimating efforts as television went from being deeply uncomfortable with its lower-class image to fully coming out by embracing the many desires and viewers it appealed to. In this episode, heterosexual and class anxieties overlap as an interest in quality television and luxury goods is marked as gay.

Yet this episode also destabilizes the dominant ideologies that frame homosexuality as a joke. The audience is encouraged to laugh at Spence's absurd homophobia and heterofailure. This failure lies not with the recommendations but rather with his inability to interpret and actively choose whether to follow them. Spence inputs several quality television series, and instead of understanding TiVo's recommendations as a list of texts associated with affluence and elitism, he construes them (and affluent culture) in a way that he rejects, or self-consciously thinks others will reject. The point is not whether Spence interprets these recommendations correctly but rather that they are queerly open to interpretation; one could see them as meaning most anything, or nothing at all. The laugh track erupts at moments that reveal how anxious Spence becomes due to these recommendations and what they perhaps reveal to both him and others concerning his own status as a heterosexual. Late in the episode, Spence tries to correct TiVo's "dumb mistakes" by recording football games, NASCAR races, and basketball games. Spence does this to let TiVo know that rather than being gay, he is actually a "sports junkie," an identity Spence assumes runs in opposition to an imagined homosexual-highbrow culture.

In the end, Spence's attempt to become reempowered by resting his ability to define himself on TiVo only results in his recording, watching, and consuming more media—TiVo's goal all along. Yet he has misunderstood the TiVo system and assumed that later actions simply nullify past (recording) transgressions, when really TiVo's algorithms work to find the intersections between such opposing programs. The result is a set of recordings like *Men's Ice Dancing* and *Breaking the Surface: The Greg Louganis Story*. Although his neighbor takes TiVo's side and intimates that it does seem to be intuitive, Spence responds with a sudden outburst: "I'm not gay. I love women, I stalk them, I'll have sex with you right now if you want." As the laugh track suddenly erupts, the neighbor leaves the apartment in a rush and Spence screams at his television, "Oh how dare you label me!" and then proceeds to call his TiVo gay. Later, Spence announces that he is being sexually harassed by TiVo.

By equating "labeling" with sexual harassment, Spence frames his anger as arising not from homophobia per se but rather from having his identity judged at all. Immediately after he tries to embrace a pathological masculinity by proudly stating that he stalks women and will have sex with a woman at the drop of a hat, he ironically uses queer and feminist rhetorical strategies to challenge not only TiVo's representation of himself as gay (or interested in gay subject matter) but also any form of representation that seeks to label people with static categories. Spence would not be angry if TiVo had presented him with a list of suggestions that had stronger heterosexual connotations (like *King of Queens* itself), not just because he thinks of heterosexuality as being a more positive category than homosexuality (which he largely thinks of as a slur) but also, more importantly, because he identifies himself as a heterosexual. These fears are made only more ironic by the name *King of Queens*, a title also shared by drag queen television star RuPaul.[63] If Spence openly identified himself as a homosexual, TiVo's labeling him as such would not upset him, or at least not for the same reasons. Instead, he might have been bothered by the broad surveillance paradigm that TiVo and technologies like it helped naturalize and domesticate—a more reasonable fear completely ignored by *King of Queens*.

Although TiVo's ability to destabilize identity in such an intimate way may suggest its queer and transgressive potential, such potential is ultimately limited and made laughable. Both *The Mind of the Married Man* and *King of Queens* present a character's panic as a momentary misunderstanding rather than as an affirmation of the complexity and instability of gender and sexual identity. These plotlines place the character's sexual identity in question only to assert—and straighten—the lines between such identities. The idea that there is a direct relationship between enjoying a series with a gay theme or a largely gay audience and necessarily being gay is, of course, far too simplistic. Both *Sex and the City* and *Six Feet Under* (as well as many of the other series mentioned) are notable not because they exclude a heterosexual audience but because they encourage a metrosexual crossover through a shared focus on affluence and empowered consumerism. Yet, though these series and TiVo's suggestions allow for a more fluid relationship between sexuality, taste, and empowerment, here these efforts backfire as Spence reacts against them in favor of traditional heteronormativity and its static binaries. This moment illustrates a push and pull between a phallogocentric view of our technologies that asserts we should become more like they see us and an impulse to reject this perspective when it contradicts how we see ourselves. In the process, recommendation systems become a space for us to consider both who we are, how others might see us, and who we would like to be.

Although *The Mind of the Married Man* and *King of Queens* are fiction, the anxiety concerning TiVo's role in imagining microcast audiences in relation to sexual and class identities is not. On November 26, 2002, Jeffrey Zaslow of the *Wall Street Journal* reported on numerous people all experiencing the same discomfort with how TiVo rendered their identities. In his article, titled "If TiVo Thinks You Are Gay, Here's How to Set It Straight," Zaslow, at once humorous and earnest, discusses instances in which real people—along with television characters—fought back against TiVo's "cocksure assumptions about them that are way off base."[64] In his article, Hollywood producers in particular voiced their concerns about privacy and sexuality. The inclusion of these perspectives indicates how TiVo's recommendations potentially could affect not only the way spectators experience their media but also how these media get made. One of the more notable executive producers Zaslow interviewed, Basil Iwanyk, described himself as the "straightest guy on Earth," though he would later ironically produce such hyperqueer opuses as *Clash of the Titans* (2010), *The Expendables* (2010), and *The Town* (2010).[65] The knowledge that powerful auteurs who have in-depth knowledge of media production use TiVo also makes the technology appear more legitimate. Zaslow likely focused in part on Hollywood executives because, in 2002, TiVo was both a luxury item priced out of reach for most people and a tax write-off for television and media producers.

The anecdotes in Zaslow's piece illustrate how the straight panic and sexual anxieties that surrounded microcasting audiences also affected (and were, in turn, reproduced by) media producers. For Iwanyk, the TiVo Suggestions Service became a thorn in his side when it started to "inexplicably" record programs with gay themes.[66] These automated recommendations seemed to be telling Iwanyk that he desired the same things as others who gave a thumbs-up to gay-themed television series—namely, other men. Iwanyk's recommendations gave him a sudden insight into an alternative way that his film interests could be interpreted that he flatly rejected. The question for Iwanyk was never whether

TiVo was queering the series he enjoyed but rather whether it was queering him. In either case, by illustrating the polysemic nature of these media products, TiVo points to how algorithmic culture and the recommendations it fosters do not just allow for reading texts against the grain in at least potentially transgressive ways but also often encourage such activities. It is tempting to think that algorithmic culture, with its focus on creating an aura of objectivity and mechanical order, would be absent of moments of serendipity, or even mere surprise, but these stories of decoding and misreading recommendations illustrate ways in which these systems continue to hold subversive potential.

And yet, much of this potential for misreading, surprise, and transgression comes from the obscured specifics behind how TiVo—and algorithmic culture more generally—generates its recommendations. Like Amazon, Netflix, Google, Facebook, and any other company that uses personalization technologies, TiVo primarily generates recommendations to users based on what similar users have liked in the past. Although this logic is intuitive, it also hides the complexity of the underlying secret and proprietary algorithms and the statistical models that govern exactly how similarities between users are defined. TiVo's automated recommendations demonstrate that there is a link between a piece of media and the users who consume it, but the qualities that unite them are rarely definable, and TiVo's software does not specify exactly why it generates the recommendations it does. When TiVo recommends *King of Queens* to viewers of *Will and Grace*, it does not explain why it finds an affinity between these shows, whether based on genre, cast, color palette, set design, or another factor. Users are left to imagine what aspects tie various recommendations together and, more importantly, what these recommendations say about themselves and their desires. This process is made more complicated by the various crossover audiences that contemporary television series try to attract.

While the concealment of the logic behind these recommendations can be threatening, it also encourages users to come up with their own readings and explanations for the recommendations. Without explanations, Iwanyk and others like him jumped to the conclusion that the quality that linked them with their recommended series was related to sexuality and that their actions were causing TiVo to represent or even "think" of them as homosexuals. These users viewed TiVo as a disempowering force that destabilized their tastes and sense of self. Since TiVo keeps its algorithms secret, there is no way to know whether or how it (or any other company) takes stereotypes into account and whether its recommendations really are meant to be based on sexuality or any other identity markers. Iwanyk's reaction is an example of how these strategies lead users to read stereotypes back into their generated recommendations. Instead of questioning TiVo, Iwanyk questioned himself and started watching, recording, and rating more television that he believed a heterosexual male would be interested in (i.e., series that he referred to as "guy stuff").[67] He seems to imagine that if the algorithm displays him as straight, then apparently he is straight. Unfortunately for him, he received suggestions like documentaries on Adolf Eichmann and Joseph Goebbels: "It stopped thinking I was gay and decided I was a crazy guy reminiscing about the Third Reich."[68] Iwanyk's reaction illustrates a confusion between gender and sexuality, as well as his assumption that homosexuality is a form of femininity. Such confusion is mapped onto and facilitated by the confusion about how secretive algorithmic recommendations really work and what kinds of explanatory or perceptive powers they might hold, if any.

All of the foregoing accounts, whether real or fictional, exemplify users' desire for TiVo to get them "right." These users felt disempowered by TiVo precisely because it did not see them as they wanted to be seen. A contrasting and more subversive reaction to TiVo and its surveillance regime, however, is the desire for it *not* to recognize the user correctly. Enacting this is unpleasant and requires users to record and rate highly things they have no interest in and stop viewing what they actually want to; indeed, such activities take away any benefits or pleasures that owning a TiVo (not to mention a television) might entail. And yet, several of the people in Zaslow's article who self-identified as gay did just that because they did not like the idea of their media both recognizing and recording their sexuality. One of the two, Ray Everett-Church, an internet privacy consultant, said his "TiVo quickly figured out that he and his partner were gay," and though this did not negatively affect them, they decided to try to "confuse the software by punching in 'redneck' programs, like Jerry Springer's talk show."[69] By recording content they were not interested in viewing, this couple tried to break the algorithm with the express purpose of regaining their agency, even though that largely defeated the purpose of owning a TiVo and denied the pleasures that might come from it. These users frame their attempts to control how they were represented by TiVo, particularly in terms of sexuality, as playfully subversive; they took TiVo up on its proposition to "program your own TV networks," but (ironically) this programming ran counter to TiVo's own recommendations.

Focusing on the question of what to do if TiVo thinks you are gay is a provocative, if ultimately vexed, way to address the complexity of subversion, misreading, and agency in digital recommendation systems and algorithmic culture more generally. The question itself, along with the reactions and texts that it generated, can and should be viewed as simultaneously subversive and docile, as it points to the blurring lines of both sexuality and digital identity, but ultimately in a way that never quite challenges the status quo of heteronormativity or digital surveillance. The responses from viewers are never to simply return their TiVos, which would potentially be much more disruptive to algorithmic culture generally, but are rather focused on manipulating the technology to reassert personal control, agency, and privacy.

Ultimately, four years later, TiVo appropriated this meme of the "gay" recommended user to associate itself more closely with gay audiences by placing the question that I have focused on here in an ad in the 2006 GLAAD (Gay and Lesbian Alliance against Defamation) Media Awards program booklet. Along with congratulating all the nominees for awards, the ad depicted a TiVo menu full of LGBT-centric programming and asked, "Does your TiVo box think you're gay? You decide. TiVo, TV your way." Although this joke implies a heterosexual audience mistaken for being gay, here TiVo reworks it to instead both affirm the potential benefits of being recognized (without judgment) as gay and laud the increasing amount of television content that features LGBT-centric characters and plotlines. Here, TiVo uses this meme to redefine itself as a tool that no longer stands in the way of user agency and privacy but instead ensures these desires, especially in relation to sexual identity. It illustrates that, like all forms of communication, algorithms are polysemic and can transform in sudden and unexpected ways.

Algorithmic Uncertainties (Netflix and the *Napoleon Dynamite* Problem)

> Every day Americans are forced to provide to businesses and others personal information without having any control over where that information goes. These records are a window into our loves, likes and dislikes. —Senator Paul Simon[70]

TiVo raised questions for users concerning the relationship between sexuality and one's place within a digital community by showing how mediated choices become possibly disturbing representations of the self. Yet it also illustrated that we often react to such representations with a camp sensibility. With this sensibility, we find agency in moments that appear disempowering.

Like TiVo's, Netflix's history presents us with numerous examples of how algorithmic recommendations can be read in any of a number of ways and reappropriated for one's own goals. While I discussed the creation of the Babadook as a gay icon in my introduction, here I discuss the Netflix Prize, a contest open to the public from 2006 to 2010 that promised to award $1 million to anyone who could improve Netflix's recommendation system by 10 percent. This competition brought to light not just how recommendations can be read by users in negotiated or oppositional ways but also how the recommendation algorithms themselves also analyze and shape the data they receive in complex, subjective ways that reveal the historicity of algorithmic culture more broadly. Founded in 1997, Netflix began its mail-delivery DVD rental service at the same time that TiVo started selling PVRs in 1999. As Kevin McDonald has argued, Netflix also heavily marketed its "Suggestion Page" as a way to indicate the premium nature of its service and separate it from competitors.[71] By 2006, Netflix was shipping 1.4 million discs per day and had amassed over 1.4 billion film ratings from its five million active customers.[72] By 2011, approximately 60 percent of Netflix users relied on these recommendations in order to decide which films to view.[73] By 2013, 75 percent of all user activity was driven by these recommendations.[74] In comparison to eBay, where "90 percent of what people buy . . . comes from search," users now, especially when streaming films, view recommendations and only resort to searches when all else fails.[75] Even when one does search for a particular film, recommendations pop up to supply alternatives just in case a particular film is unavailable or the user changes his or her mind.

Netflix has often cited its recommender software, called Cinematch, as being central to its popularity and growth. During the height of its pre-streaming market in the late 2000s and early 2010s, when its library included over one hundred thousand films and television series available for rent, Netflix needed to help users find the media they might like best. Reed Hastings, the president of Netflix, has said that this technology is necessary to guide Netflix's subscribers through the difficulty of making a choice when confronted by an overwhelming number of possibilities.[76] Cinematch was central to Netflix's advertising throughout its early years as the company lured people away from their local video rental stores, where they could easily browse through aisles of videos or ask a clerk for a recommendation. Cinematch replaces a clerk's recommendation in a space that features tens of thousands more viewing possibilities. It has largely been successful in this effort as, while Cinematch has improved, so has the number of positively rated films on the site, showing that people now generally rent films on Netflix that they end up enjoying.[77] In 2006, Netflix reported that users view twice as many films as they did before joining the service and together rated about four million films a day.[78] Through these ratings, which in 2011 numbered over five billion, Netflix supplied its twenty-three million members with recommendations concerning what else they should watch.[79] By 2015, the number of global subscribers had grown to 57.4 million, and by 2018 that number doubled to nearly 118 million.[80]

One important result of the use of Cinematch along with Netflix's subscription-based business model is that, in 2006, while 80 percent of the films rented in stores were blockbuster new releases, this statistic was nearly reversed for Netflix. On this site, 70 percent of the films rented were backlist titles, meaning that they were older films, foreign, or less popular new releases.[81] With new releases often costing more to license and having more strings attached to their rentals, Netflix continues to have a large incentive for trying to instead steer its customer base toward backlist titles. Not just an impetus for the recommendation system itself, it also structures how and what Cinematch recommends.

While Netflix and other companies generally keep secret the exact ways by which their proprietary digital recommendation systems operate, in October 2006, Netflix announced the Netflix Prize contest, which helped expose to the public the various technologies and assumptions on which these systems are based. At this point, Netflix was unable to markedly improve Cinematch's accuracy and feared the costs of hiring more software engineers who, in the end, might not be able to help it. Instead, it asked the public for help and offered $1 million as a prize to anyone who could improve its Cinematch recommendations by 10 percent. To help contestants test their algorithms, Netflix released to the public a database containing one hundred million ratings from 480,000 of its customers, covering 1,800 movie titles.[82] This data set, which was anonymized and "perturbed" (slightly altered through statistical means) to protect users' confidentiality, was the largest of its kind ever made available to the public.[83]

Even at $1 million, the Netflix Prize was cheap in comparison to the salaries the company would have otherwise paid to full-time programmers and also resulted in a great deal of free advertising. In a neoliberal economy that has forced many into low-paid and precarious labor, the Netflix Prize became a model for how companies could outsource even their most complicated and technical dilemmas. Rather than criticize Netflix for not hiring more experts in the field, business and technology blogs and magazines universally praised it for its creative solution and its ability to get hundreds of extremely intelligent people with a wide variety of skill sets across the globe to effectively work together for free on a problem that might not even have had a solution.[84] Many of the people working on this quandary labored on similar issues in their actual jobs and spent their free time trying to apply their knowledge to win the prize. As John Thornton Caldwell, Mark Deuze, and I elsewhere have argued, prizes have long been used to get people to contribute their unpaid labor and time to ultimately help a company make more money, and Netflix has become the primary example of how this logic can be applied to the much larger audiences of the World Wide Web in ways that make contestants happy even as they lessen the value of technical labor for everyone.[85]

In September 2009, Netflix awarded the $1 million grand prize to the nine international members of the winning team, BellKor's Pragmatic Chaos. As many of the contestants were academics and computer science researchers, there are conference and journal articles that chronicle both their successes and their failures. As Blake Hallinan and Ted Striphas demonstrate, these articles help to sketch out both the standard methods for rating films and the many theoretical and philosophical questions that must be addressed by such efforts, including how these efforts change our very conception of culture.[86] While filmmakers, theorists, and video store clerks argue that audiences enjoy films based on genre or the people involved in their production, collaborative filtering offers a very different and much more ineffable vision of how film taste works. Collaborative filtering can lead to obvious pairings. Netflix might see that you liked *Nightmare on Elm Street* and *Friday the 13th*. It might also see that others who liked those two films also liked *Children of the Corn* and therefore recommend it to you. But other pairings are much less obvious. For instance, Martin Chabbert, one Netflix Prize contestant, used a particular collaborative filtering method called singular value decomposition to discover that, based on a large number of ratings from many users, there is a particular quality that unites "a historical movie, 'Joan of Arc,' a wrestling video, 'W.W.E.: SummerSlam 2004,' the comedy 'It Had to Be You' and a version of Charles Dickens's 'Bleak House.'"[87] While neither these statistical formulas nor any contemporary theoretical framework can help us to figure out what this similarity consists of or why a person who enjoys *Joan of Arc* and *Bleak House* would also likely find pleasure in the *WWE: SummerSlam 2004*, Chabbert's algorithms showed that this is in fact the case. Yet by assuming that Chabbert's findings are correct even though they make no sense, we may be discounting the very real possibility that the algorithm might simply be wrong or that it is not discovering a connection but rather creating it. The underlying assumption behind this way of creating recommendation systems is that while everyone's taste may be different, the way we as humans individually come to our taste is essentially biological and universal.

Other contestants focused on questions of similarity and pondered how many sequels, remakes, seasons of a series, and rehashed scripts a viewer would enjoy before becoming fatigued. Contestants considered capricious aspects like the time of day, the weather forecast, and a user's dating status for how those factors might affect the enjoyment of a film. These contingencies complicated the issue that users' opinions of films and their concurrent ratings tended to change by about half a star within just a month. Still others concentrated on those polarizing films that do not seem to fit into any clear genre or category and that are therefore very difficult to recommend accurately; this was referred to as the "Napoleon Dynamite" problem because, as demonstrated by the response to *Napoleon Dynamite*, there seemed to be no obvious predictors for whether a user would enjoy such a film.[88]

These various foci illustrate not only how many details can affect our interest in media but also how prioritizing certain details over (or instead of) others can result in very different conceptions of taste. Yet here we can also think of *Napoleon Dynamite* as approaching a Derridean trace, or a pointer that illustrates the instability of algorithmic "knowledge." By being *un*recommendable, it disrupts the very notion that algorithms can ever really be accurate or know us as more than a vague approximation. Rather than challenge the very premise of the Netflix Prize, contestants instead doubled down on their assumption that more and different data was all that was necessary. Some tried to integrate user information from social networking sites into their recommendations, believing that knowing more about a person and his or her community would be helpful in divining the individual's filmic taste.[89] Others tried to use qualitative information from Wikipedia about each film to make recommendations, with the belief that a person who liked one particular kind of film would continue to like other films of that particular type.[90] Surprisingly, neither information about a film nor that about a user from a social networking site improved Netflix's existing system, which, at that point, only took into account ratings created by users.[91] Instead, the winning method by BellKor's Pragmatic Chaos improved on the system Netflix already had in place by analyzing the user data with a much larger number of collaborative filtering algorithms (each of which has its own strengths and weaknesses) and finding the most "accurate" weighted average of all of their computed ratings.

Yet, in the end, after paying $1 million to BellKor's Pragmatic Chaos, Netflix ultimately decided not to employ the improvements. On the *Netflix Tech Blog*, engineers Xavier Amatriain and Justin Basilico reported that while they had integrated some early promising algorithms from the prize into Cinematch, they felt that it was not worth the time to implement the winning method.[92] At the time, Netflix was slowly moving from U.S. DVD distribution to a global streaming service, and they argued that the DVD-based five-star rating system did not work well for streaming. They also stated that streaming was bringing them a lot more data than just a star rating, such as whether users actually finished films, how long it took them to watch complete series, whether they took breaks, when they viewed certain types of films, and so on. They argued not only that all of this information could be used to create recommendations but also that they were moving toward an interface in which every item users see (and the order those items are presented in) is chosen by their recommendation system.

Finn argues that this represented a "reinvention of the Netflix culture machine" wherein "all of a sudden, cultural reality leaks into the statistical cleanroom where algorithms count nothing but users, movies, ratings, and timestamps."[93] He points out that with this change, Netflix began moving toward creating recommendations based on ever more pieces of metadata gathered both automatically from users and manually by paid viewers who would tag movies with various keywords. While Finn uses this moment to explore "the function that magic"—or, perhaps, Logos—"still plays in computation as a way to bridge causal gaps in complex systems," I would argue that it also illustrates a turning away from a purely phallogocentric way of defining recommendations and our relationship to computers more generally.[94] While Netflix is certainly going all in on the idea that more data will result in better recommendations, it also states that no one recommendation system will fit all people all of the time. Here, it begins to allow for the idea that not everyone wants to watch

movies just because they might enjoy them: some want to see popular or critically acclaimed ones even if they might not like them. It is perhaps not a
coincidence that the term *hate watching* was coined right around the same
time. Amatriain and Basilico also present a long list of recommendation algorithms that each, they argue, have their positives and negatives.[95] Rather than
assert that they will ever be able to find a perfectly accurate algorithm, they
use this long list to illustrate that Netflix must continually change and improve
the ones it has both because not all people watch media for the same reasons
and because people change.

At the same time, this change was perhaps made for other reasons entirely.
Amatriain and Basilico present the rational reasons for changing Netflix's recommendations, but there are also clear economic reasons for changing them
as well. As I discussed in relation to *House of Cards*, Netflix was beginning to
exhibit and produce its own content at this point and may have wanted to
change how it recommended titles to customers in order to justify recommending all Netflix productions to all of its users. With the previous model, it
would have needed a personal reason for recommending its series to customers, but with its new and more open model, it could recommend all of its series
to every customer, regardless of whether it had any reason to think they might
like them.

Christian Sandvig has called this type of practice "corrupt personalization,"
as it presents general recommendations (that may actually be at odds with our
interests) as personalized ones.[96] While I agree that it would be nice if corporations more transparently explained and labeled how their algorithms work,
users must be a bit more skeptical of the algorithms that surround them. I would
also caution against the idea that an algorithm or a recommendation could ever
be purely benevolent or corrupt. As the Netflix Prize and its aftermath illustrate, algorithms are extremely complex and are created as an assemblage of
multiple concerns, perspectives, and goals. As industries and their goals (or
the cultures around them) transform, so do their algorithms. This moment
shows the impossibility of any project that aims to create a recommendation
system that may accurately predict the taste of all users and illustrates that,
instead, algorithms are continually updated with no expectation of ever
reaching perfection.

Algorithmic Subversions (Breaking Digg's Axle)

Just as in the case of other media and forms of entertainment, the news has long
been algorithmically sorted and recommended based on what users might most
want to see rather than what may be most important; these algorithms have
turned the news into an entertainment much like any other. Here, I will discuss how this process shaped early conceptions of digital citizenship and

empowerment on Digg, an early news and media aggregator wherein users could vote on and recommend content they felt should be visible to a larger public. The creators of Digg tried to organize its ideals of digital citizenship around a celebration of its seemingly fair and objective collaborative filters and ranking systems, which they felt could organize an ideal digital culture. Yet here I discuss how the user base ultimately rebelled against Digg's algorithmic righteousness and reasserted its agency and critical consciousness.

From its founding in 2004 to its change of ownership in 2012, Digg employed ostensibly democratic technologies and rhetoric by asking users to submit news and entertainment stories and vote on which deserved the public's attention. This site became popular with users interested in discovering, distributing, and discussing online content that mainstream news and entertainment presses had skipped over. Communities developed on the site around a shared desire to shape public discourse by making this content (whether serious stories on political candidates, discussions on new technologies, or entertaining videos of kittens) more visible to a global audience. Those who submitted stories that became prominent also became more influential and powerful as Digg users and as part of the Digg community.

Kevin Rose, a cofounder and chief architect of Digg, repeatedly declared that he wanted the site to be a "true, free, democratic social platform devoid of monetary motivations" that would give power back to the "masses."[97] While it eventually dropped in popularity and in 2012 was broken up, sold, and completely redesigned by its new owner, a venture capital company called Beta-Works, during its heyday from 2006 to 2010 (the period I will focus on), Digg had millions of unique monthly viewers and was one of the top 150 visited sites globally and one of the top 100 in the United States.[98]

Yet, at the same time, Digg's algorithms were designed not just to raise important stories to the public's attention but also to make fallacious ones disappear. While the 2016 U.S. presidential election put a spotlight on the rapidly growing global economy and technologies of fake news, I hope my focus on Digg illustrates the longer history of this phenomenon and the role that recommendation systems have played in both spreading and confronting such content. As people across the globe, including a group of entrepreneurial Macedonian teenagers, wrote and posted salacious and fallacious articles on Hillary Clinton, Donald Trump, and various other political figures, many across the political spectrum were duped and shocked (shocked!) to discover how prevalent deception had become online.[99] Much of the resulting surprise and anger illustrated a paradoxical belief that, on the one hand, due to its billions of algorithms and users, the internet is fundamentally a public good and should be able to self-correct and stamp out all lies before they reach us; James Surowiecki refers to this utopian belief as "the wisdom of crowds."[100] This hope that the internet and algorithms could simply be a positive democratizing tool for the empowerment of well-meaning and rational citizens led to the minimization of the more nefarious, nihilistic, and anarchic elements that make both the internet and humanity so complex. On the other hand, due to this huge amount of information, we also believe that we are incapable of adequately understanding, let alone critiquing, all of the information that comes at us; the only thing capable of judging the internet often seems to be the internet itself.

Debates have long been waged over whether the World Wide Web is a democratic force. In contrast to Surowiecki, Jodi Dean argues that sites like Digg "present themselves for and as a democratic public" in ways that conflate democracy with the "eager offering of information, access, and opportunity" and with the "expansions in the infrastructure of the information society."[101] As democracy becomes equated simply with networking, equity and diversity become associated with the desired increase and commodification of digital content rather than the reverence of all human life. Instead of democracy, these technologies yield "communicative capitalism," a system that presents every part of life as a form of spectacle and publicity for the market that "undermines political opportunity and efficacy for most of the world's people."[102] Yet one may ask whether the practice of democracy was ever so pure and whether it has always been a fantasy designed to rationalize systemic inequities and injustices.

Much like Finn's, Cheney-Lippold's, and Pasquale's, Dean's description of communicative capitalism assumes that most humans have virtually no agency or ability to question or combat the harmful effects of algorithms and the culture they create for us. While Rose continually presented Digg through a rhetoric of community and democracy, this message was often at odds with the venture capitalist funding that kept the site running but also wanted to see it eventually turn a profit. Here, I analyze how various groups were in disagreement over how Digg's digital recommendation and voting system should function and what kind of model of community it should enable. The site's creators and users both identify the recommendation as the central logic of digital democracy practices and furthermore indicate how this logic figured values of equity and diversity in often contradictory ways. Digg's recommendation system, its most unique and influential characteristic, allowed users to either "digg" a story, thereby declaring that it was interesting and should be noticed by others, or "bury" it, thus asserting that the story was spam, derogatory, fallacious, boring, or otherwise not worth reading or viewing. Rather than employ editors or producers to filter and rank content, Digg's recommendation system ranked and "gatekept" content based solely on these thumbs-up and thumbs-down votes. The more popular a story became, the higher up the list it moved, until it arrived on the front page of Digg, where it could then become noticeable to the highest number of people.

Digg was one of many companies, including Slashdot and Reddit, working on the question of how to sift through the abundance of web content in ways that were cheap, high quality, and potentially profitable. Axel Bruns argues that the web ushered in a sudden change to editing practices and publishing industries, as now everyone had the potential to be a publisher of content. He asks, "If audiences have instant access to a myriad of information sources, what factors determine their choices, and how do their choices affect their views and knowledge of the world around them?"[103] Digg's recommendation system was an influential answer to this question, as it worked to shape the choices of audiences in ways that helped stabilize and publicize the conventional publishing industry, a process that Bruns refers to as gatewatching. Indeed, its recommendations took choice out of the equation by making the web appear more like a short, curated list of articles rather than an endless abundance of unsorted data. The automated nature of this gatewatching appears scientifically objective and therefore also equitable, or fair and impartial. Yet the same technologies that allowed users to rank content also made it possible for Digg to rank users and treat them differently by making their votes worth more or less depending on how much time, labor, and thereby value they had contributed to the site; as a result, the diversity of content and users on the site suffered.

Before 2010, Digg used these rankings to reward those who uploaded and dug more with greater influence. In the process, they generated a meritocracy rather than a democracy. While highly ranked users, who often spent a great deal of time searching for and uploading content to Digg, found this system fair and equitable, lower-ranked users, who were often newer to and not yet as enmeshed in the Digg community, often expressed frustration with their inability to get stories noticed by Digg users as a whole.[104] After 2010, in an effort to appeal to newer and wider audiences, Digg algorithmically limited the influence of those with high ranks to make it easier for more users to get their uploaded content widely seen. At the same time, it also employed celebratory democratic rhetoric to encourage users to upload more, vote more, and otherwise work harder to make Digg successful by framing the site as an equitable community of users rather than a top-down business. While Digg made these changes to its recommendation system to make the site more "democratic" and attract a set of users with more diverse interests, many long-time users felt that it also led to a reduction of both the equity of users and the diversity of content.

The algorithms Digg used to rank its content from 2006 to 2010 structured how its user community conceived of democracy within a digital space in ways that opposed the values of diversity and equity. As Matthew Hindman has noted, digital democracy is often unthinkingly praised because the term *democracy* itself lacks specificity and is often simply defined as something that can be praised: "To say that the Internet is a democratic technology is to say that it is a good thing."[105] Diversity and equity are largely imagined to be part of the democratic (good) project; yet these values get folded into a phallogocentric project that assumes that an algorithm can effectively automate them for all. Yet Hindman fetishistically notes that while the internet may hold the promise that many more people will be able to state their ideas in a public forum, it is less clear whether these ideas will be taken seriously, lead to more diversity (of content or citizens), or even be heard at all. Indeed, sites that encourage users to contribute and discuss content are often dominated by a small number of users who wield large amounts of influence and power. Financial journalist John Gapper notes that these unpaid users who spend a great deal of time on these sites often "achieve an elite status reminiscent of old media's professional gatekeepers."[106] These users may generate a diversity of content, but this does not imply that the sites have or celebrate a diverse user base or an equitable set of governing rules or culture.

While Digg often used the rhetoric of democracy, diversity, and equity when discussing its technologies and user base, it also ran into continual problems when designing its website with such values in mind. At the time of its founding, American political campaigns were beginning to rely heavily on digital technologies to create and spread grassroots support in ways that commentators framed as a great step for democracy. In discussing Howard Dean's 2004 presidential campaign, Lee Rainie, John B. Horrigan, and Michael Cornfield point out that the relatively new trends of blogging and meetups (among others) helped to bypass mainstream media and gave the electorate more power.[107] Ian Bogost argues further that social networking and other technologies that united large groups of people in virtual communities were also an important part of this burgeoning "digital democracy."[108] Yet he also argues that none of these technologies actually worked to embody the logic of or give users an experience of democracy. Indeed, Bogost calls for the development of more digital technologies that use their underlying algorithmic rules (or procedures) to mimic the rules of democracy in order to give users a better understanding of the political process.

While the exact mechanisms of Digg's voting and recommendation algorithms are proprietary and have not been released to the public, they were presented as efforts to embody the neoliberal logic of democracy, collaboration, and personalization. Yet the rhetoric of Digg's owners and users also illuminates how collaborative, democratic rhetoric can be used to disempower users. In order to encourage users to visit its site and thus make it more marketable, Digg presented the labor of users as empowering communities. Rose referred to Digg users as the "DiggNation" and relied on the site's algorithms to give users a stake in the site's success. On August 25, 2010, however, Digg revised its recommendation algorithms and the way it counted and weighed the votes of users. The expressed motive was to make the site open to more people. Yet these changes resulted in a massive revolt from its community and huge debates over the relative value of different types of recommendation systems and the values they seemed to instill. After the changes to Digg's recommendation technologies in 2010, users debated whether the old algorithms had ranked content in a "fair" way that accurately reflected the will of Digg's users as a whole, or whether they rather displayed content that only a few users preferred.

In 2010, before the change in algorithms, the creators of the site described it as follows:

> Digg is a place for people to discover and share content from anywhere on the web. From the biggest online destinations to the most obscure blog, Digg surfaces the best stuff as voted on by our users. You won't find editors at Digg—we're here to provide a place where people can collectively determine the value of content and we're changing the way people consume information online. How do we do this? Everything on Digg—from news to videos to images—is submitted by our community. . . . Because Digg is all about sharing and discovery, there's a conversation that happens around the content. We're here to promote that conversation and provide tools for our community to discuss the topics that they're passionate about. By looking at information through the lens of the collective community on Digg, you'll always find something interesting and unique. We're committed to giving every piece of content on the web an equal shot at being the next big thing.[109]

This description, laden with tech industry buzzwords, focuses on Digg's interest in creating a collaborative, connected, and collective community empowered to make stories that mainstream news media may have missed. It frames Digg as a neutral, passive, and benevolent platform that only echoes the beliefs and ideas of its users. This phrasing obscures the fact that voting and recommendation systems, and algorithms more generally, can shape discourse and offer up a politics of their own. Recommendations, whether built on democratic, egalitarian, or hierarchical models, encourage certain discoveries, discussions, and communities over others. Digg's collaborative filtering algorithms, rules, and systems—and the way they bring certain stories into the public eye while neglecting others—are largely proprietary and invisible to most users. As Tarleton Gillespie has argued, such elisions make platforms more appealing to a wider audience that may include members with various beliefs, and they also help protect sites from being responsible for the content members add to them.[110] Before the change in algorithms, Rose stated that the Digg algorithm treated all users equally and that specific viewpoints and stories became more prominent only by the concerted effort of the community as a whole.[111] While Rose asserts that Digg made the site's algorithm purposefully opaque and complicated only to ensure that users could not easily manipulate it, this lack of transparency makes the site falsely appear to be value neutral.

Indeed, users quickly found ways to exploit Digg's algorithms to get their stories more publicity. In 2007 Muhammad Saleem, a prominent social media blogger and Digg user, asserted that Digg did not require a specific number or percentage of votes for a story to reach the front page.[112] Rather, Digg's algorithm took into account the recent participation levels of users and their followers (those who have requested to be alerted when specific users contribute content) in deciding how easy it would be for them to get stories onto the front page. Along with participation, if many "high-value" users (known as "power diggers") quickly dugg a story, it would take far fewer votes for it to reach the front page.[113]

This understanding drove the creation of particular kinds of voting bloc communities on Digg that voted up each other's stories, regardless of quality, in order to make their votes more influential. This also led to Digg's featuring the perspectives of these users more heavily than others. From 2006 to 2008, while Digg asserted that it gave each user an equal say on the site, its proprietary collaborative-filtering-based voting system made the votes of prolific users worth more than the votes of those who rarely interacted with the site. Rand Fishkin, a Digg user and creator of Moz, an online marketing firm, argues that, "far from being a mass of opinion, Digg is instead showing, primarily, the content opinions of just a few, select folks."[114] Rather than a democracy, Digg operated as a "frathouse" where it paid "to know the right people."[115] Digg's voting system made certain users more equal than others and created what Richard Adhikari at TechNewsWorld refers to as a "tyranny of the minority."[116]

In 2006 the top one hundred Digg users were responsible for submitting 56 percent of all stories. In addition, the top twenty users contributed nearly 21 percent of the 25,260 stories that reached Digg's front page. While this number dropped slightly in 2007, these rarefied power diggers continued to dominate the site, with the top one hundred submitting 44 percent of all stories.[117] The top digger in 2007 was responsible for 3 percent of all front-page stories.[118] The extra weight Digg's voting system gave to these users, as well as the tit-for-tat relationships that developed within this community, created this extreme imbalance. Power diggers spent many hours on this site and developed friendships with other users, enabling them to both chat and follow each other's diggs. Members in these groups help each other out by digging each other's articles. As a result, those who were well connected to the larger community were rewarded by having more influence over which stories made it to the front page to be seen by the most readers and potentially also other mainstream media outlets.

Even as Digg's algorithms continued to value the voices of a small minority of users above all others, Rose continually celebrated its "democratic" objective of allowing users, regardless of their wealth, fame, or power, to influence the national dialogue. Rose made a video in 2008 emphatically stating that Digg is a democracy that views its users as equals. Many also reported on how Digg's unique status as a news site where the users are the editors stems from democratic aspirations.[119] Rose hosted a weekly podcast from 2005 to 2012 called *Diggnation*, in which he and others discussed those articles that were currently popular on Digg—while imbibing and advertising a wide assortment of alcohols and iced teas. Users often employed a similar rhetoric when describing Digg and their loyalty to it. Along with being called the "Digg Nation," they referred to themselves variously as "brigades," "patriots," and "rebels."[120] Users also accused Digg of creating an "army" devoted to promoting certain stories over others.[121]

Digg's use and encouragement of this nationalistic rhetoric shows how the catchphrases of digital democracy and collaborative culture were used to create a community of users willing to both labor and consume to make the site a success. While user contributions are a form of work, Digg used discourses of nationalism and equity to assert that this work was a form of altruism designed to improve the community and the lives of users rather than the value of the site itself and the economic prosperity of its owners. Digg and its users thus conflated democracy and community with the economic model of the corporation. This process echoes Dean's sentiment that democracy has itself become an alias for neoliberal governance that furthers the exploitation of all labor.[122]

Although Digg's rhetoric and voting system encouraged loyalty, this system also destabilized control of the site itself by making it difficult for the its owners to dictate the types of content welcome on the site. Digg attempted to use its voting system to create an invisible infrastructure and hierarchy on the site to control its users. Yet, while recommendation and voting systems are often meant to give users the sense that website content is merely mirroring their own interests, Digg's system often reinforced the distance between Digg's users and producers. Indeed, users at times used the system to demonstrate against the will of Digg's producers. For instance, users asserted their power by forcing Digg's hand when the site appeared too focused on protecting its corporate interests. One of the most virulent revolts against Digg occurred on May 1, 2007, when a user posted an encryption key that made it possible to make copies of Blu-Rays and HD-DVDs. Fearing legal recourse, Digg took down the post and banned the user and several others who promoted it. In response, many power diggers began contributing and promoting thousands of stories that had the encryption key in their titles or were otherwise direct responses to Digg's actions. While many simply renamed articles with titles like "My name is '09-F9-11-02-9D-74-E3-5B-D8-41-56-C5-63-56-88-C0,'" one user changed Digg's Wikipedia entry to list May 1, 2007, as "the day that Digg died."[123] As the HD-DVD promotion group had advertised recently on the *Diggnation* podcast, users speculated that Digg was deleting these stories not because of legal fears but instead purely for commercial reasons. These users effectively gamed Digg's system against itself.

This episode illustrated that power diggers conceived of themselves not simply as "empowered consumers" but rather as actual producers of the site with a vote concerning both the site's content and the way the site operated. They were not in partnership but rather in contest with the actual owners of the site. Technology blogger Tamar Weinberg touted this event as a "democratic milestone" that illustrated the citizen-like power of users on sites that depend on collaborative communities.[124] Technology blogger Ryan Block highlighted the stakes of this event that pitted Digg's democratic ideals against its corporate economics when he asked, "But how did such a loyal userbase as Digg's so quickly divert its all-consuming energy to defying—even damaging—the company to which it was so loyal?"[125] In terms of content, Digg users felt that their votes made them equal not just to other users but also to the owners and operators of the site. By blocking content without the permission of users, Digg exerted an authoritarian, corporate level of control over the content on the site and threatened its democratic aura that users had worked to develop.

After witnessing the user outcry, Rose relented and unblocked all of the posts that had included the HD-DVD and Blu-Ray code in them. On Digg's blog, Rose argued that Digg had always blocked unlawful and harmful content and compared the blocking of the code to Digg's regular blocking of pornography, racial hate sites, and illegal downloads. Rose irately stated that after "thousands of comments, you've made it clear. You'd rather see Digg go down fighting than bow down to a bigger company. We hear you, and effective immediately we won't delete stories or comments containing the code and will deal with whatever the consequences might be."[126] By simultaneously celebrating the users' democratic idealism and asserting that their anticommercial spirit could be the downfall of the site, Rose illustrated the thin line Digg's owners tried to navigate between their democratic branding and corporate interests.

This conflation of democracy and corporate governance on Digg also resulted in numerous instances of Digg Nation rebellion. During the period from 2006 to 2010, when it was most popular, there were many instances in which Digg's owners weighed their responsibilities to their shareholders against their desire to keep their users (especially power diggers) happy. While they created the Digg platform, they continually suggested that they had very little control over it. The resulting conflicts concerning the interrelated status of user rights, powers, diversity, and equity often revolved around changes in Digg's voting system that affected how Digg privileges worked in relation to corporate interests. In 2008 Digg's owners tried to introduce algorithm changes and other efforts designed to increase the number, and hence the diversity, of contributors. While this sounds like a noble goal, Digg wanted more contributors partly to increase the commercial and mainstream content on Digg's front page, with the expectation that more contributors would lead to more advertising income. Power diggers pushed back because such changes reduced their sense of equity on the site by limiting their influence and ability to control content on the front page. Before its algorithmic changes, Digg gave power diggers influence at the cost of creating a diverse contributor pool. Yet Digg incentivized these users to upload articles from a diverse set of perspectives in order to keep their influence. Power diggers constructed and deconstructed coalitions constantly in order to get particular, often fringe, types of stories and messages seen by a broader audience. Such was the case in late 2007 when a group calling themselves the Bury Brigade buried positive articles on any presidential candidate who was not Ron Paul.[127] While the Bury Brigade added diversity to Digg's front page by adding their libertarian perspective to a news cycle that too often only focused on Barack Obama and John McCain, this was not the kind of diversity that Digg, the company, desired. Rather than viewing the Bury Brigade's contributions as an expression of an alternative viewpoint, Digg's owners and the larger community referred to them as spam.[128] Digg programmers saw this and similar collaborative activities as efforts to game their voting system in order to squelch the site's broad popularity and, hence, profitability.

In an effort to increase the appeal of its site to a broader swath of users, Digg increased the weight of votes cast by new and less active users (whose voices may not have been heard before on the site). On January 22, 2008, it modified its voting system to incorporate what it called a "diversity algorithm."[129] Rather than imagining diversity as a plethora of ethnicities, genders, and races, its diversity algorithm made it so that for a story to gain prominence, it had to be dug by a wide range of users who had each previously presented an interest in articles of different types and perspectives. With this change, Digg transformed itself from a site that favored the creation and proliferation of special interest groups to one that encouraged users to take an interest in a wide number of topics.

Rose announced this change on Digg's blog, stating that this "tweak" made it so that a more diverse set of users would be needed to get a story to the front page, and that this change was in line with Digg's longtime goal of both creating a "diverse, unique group of diggers" and giving "each person a fair chance of getting their submission promoted to the home page."[130] Yet there is no real sense from Rose's statement of how diversity and fairness were being defined or what about the earlier iteration of the site was less diverse or fair. Instead, Rose faintly echoed a larger anti–affirmative action discourse that uses diversity and fairness to maintain the status quo and celebrate dominant values. Importantly, this so-called diversity algorithm did not necessarily require a diverse set of *users* to digg a submission for it to become popular, as it also gave more influence to individuals who had previously dug stories on a diverse set of *topics*.[131] As the diversity algorithm continued to weigh the votes of different users in obscure ways, here *fair* and *equitable* became associated with the venture capital ideal of generating more users and more uploads. Indeed, the earlier algorithm may have been more fair and equitable in terms of rewarding users for their work.

Power diggers saw this change as a direct attack against them and their power. After the voting system's modification, these diggers immediately noticed that it was more difficult for their contributions to make it to the front page. Power digger Saleem complained that while, before the algorithm change, his stories would make it to the front page after approximately 130 votes, afterward, it took more than 200.[132] In contrast, new users were now able to get stories to the front page with only 40 votes. Indeed, Tad Hogg and Kirstina Lerman argue that these changes did result in more people getting their stories promoted to the front page, but this did not necessarily mean that the diversity of users actually shifted, because these added users may have been very similar to those who had already been successful diggers.[133] Furthermore, in a space where the physical identity of users is rarely known, diversity comes to mean something much more nebulous and its value is less certain. Power digger MrBabyMan asserted that Digg's change was a "fairly transparent strategy to clean house of the submitters who have been dominating the front page for a while now. Essentially [it] adjusted the diversity factor to skew against popular submitters."[134] Quotes like this framed the diversity algorithm as a tool designed to wrestle power and influence from user-workers and hand it back to the site's owners. While terms like *produser, prosumer*, and *playbor* have become important keywords for describing the breakdown of producers and users (and owners and workers), there are still obvious differences and distances between these two groups that Web 2.0 does not overcome. Digg's changes angered many dedicated users who felt that since they had been digging for longer, they had more expertise in finding important stories, had put in more labor, and therefore should be given preferential treatment.

The back-and-forth struggle between the owners and users of Digg came to a head on August 25, 2010, when the owners of Digg updated their site from version 3 to version 4.0, with several notable changes to their voting system. These changes most noticeably consisted of getting rid of features including an easy way to see what friends had submitted to the site; making a variety of alterations to the searching and browsing mechanism; getting rid of subcategories and certain types of searches; dispensing with the area where users could store their favorite stories; and, perhaps most important, removing the bury button and with it any way for diggers to meaningfully regulate content on the site, whether it be spam or pieces that the users simply did not like. From looking at comment boards shortly after the changes, it is clear that what upset the users the most were the changes made to the voting system and the ability of users to follow the digging activities of others.[135] Voting and keeping track of other

FIG. 3.2 Screenshot of Digg's 404 page

users were the two main functions on the site that allowed users to contribute to and collaborate with the larger Digg community. Diggers used these tools to create friendships and communities. By getting rid of them, Digg made the site a much more individualized experience, which in turn put more power in the hands of Digg's owners. Digg also gave corporations the ability to bypass users altogether and upload their own stories and advertising to the site themselves. The disappearance of the bury button and the emergence of stories contributed by corporations very clearly eliminated users' ability to control their experience of the site, disempowering their relationship to the technology.

Digg also completely revamped its MyNews page to make it friendlier to advertisers. In Digg 4.0, when users first signed up on the site, they were greeted with a page of recommended newsfeeds, including ones from the *New York Times*, the *Huffington Post*, Fox News, *Playboy*, Cracked.com, *Wired*, and other major news, entertainment, technology, and softcore porn sites. Digg's MyNews page became the default page and, rather than showing what was popular overall, it only showed stories that had been uploaded by businesses, friends, and others that many users were following. While the Digg owners' uneasy relationship with the democratic and commercial aspects of the site perpetually antagonized users, the ostentatious and sudden preferential focus on commercial perspectives in version 4.0 seriously rankled many on the comment boards who felt that this and many of the other changes were made to appease the site's advertisers and venture capitalists at the expense of the users themselves.

To make matters worse, Digg 4.0 was also extremely buggy. While many of the issues were solved relatively quickly, during the first week of the transition, many pages would take a long time to load or not load at all. Rather than display a simple 404 page to show that an error had occurred, Digg displayed an

8-bit image of the covered wagon from the 1974 *Oregon Trail* video game by MECC (the Minnesota Educational Computing Consortium) and Digg's version of the game's infamous announcement: "Digg has broken an axle. We might have to sell some oxen but we'll be back on the trail soon." Meant to provide some nostalgic and light-hearted levity to the site's technological breakdown, it also pointed again to Digg's nationalistic rhetoric. It displayed Digg and the Digg Nation as pioneers attempting to achieve their manifest destiny at the risk of losing many of their members to dysentery in the process.

Digg's update made it clear that the site's owners were more interested in creating relationships with advertisers than in creating a community of users. This change resulted in a rather sudden loss of over 25 percent of its user traffic in the United States and 35 percent in the United Kingdom, lowering it from around the 45th most popular website to approximately the 150th.[136] Perhaps even more meaningfully, sites like TechRadar that had previously gotten a huge amount of traffic from stories posted on Digg saw this traffic decrease by over 86 percent after the change, indicating that Digg had become a significantly less consequential news destination.[137] This story made its way to Digg's front page, and many diggers used the comment board to discuss their sadness over the end of Digg and to announce that they, too, were signing off for good. One digger, theKipper, stated he was closing his account because the site had become a "crappy RSS feed" with no "community, no conversation, and no feeling of human input."[138] While sad to give up on this site that they had become "emotionally attached" to, theKipper and other diggers were angered by how it was "turning corporate" and at the same time neglecting the site's actual users.

The users responded to these changes via various forms of protest, including a boycott; a "quit Digg day"; an effort to upload hundreds of stories onto Digg that would direct traffic toward Reddit, Digg's main competitor; and a great deal of aggressive behavior on the site's comment boards. What was most noticeable during the months after this change was that while beforehand the top stories on the site were viewed and made popular by thousands and tens of thousands of users, after the change, a site could easily become prominent with fewer than a hundred votes, and often fewer than twenty. Months later, after many of the changes that the user base originally rebelled against were revoked, the top story on Digg only garnered 523 votes, whereas before it would have often received somewhere between 5,000 and 20,000 votes.[139] What this change indicates is that the 25 percent of users who emigrated from Digg were those who were the most active on the site, those who voted the most and were the most dedicated to the creation and furtherance of the larger Digg community. While every vote on Digg counted, this 25 percent appeared to matter just a bit more than the rest. As one apocalyptic digger put it during this mass exodus, "The implosion of Digg is happening. It is an unstoppable conclusion to events that have been in motion since last week and will continue until the full implosion event similar to the final shrinking of the universe."[140]

Users also voiced their fears that, along with these more obvious and transparent changes, the site's very structure and voting mechanism had also been transformed to give corporations more rights than actual diggers. Through these critiques, they pointed to many of the more authoritarian aspects of Digg's attempts to make its algorithms appear to perfectly articulate its users' collective will. In response to an article concerning whether Digg gamed "its own system to benefit publisher partners," digger endersgame posted that there were many pro-Digg articles lately making it on the front page with few votes and all the ones complaining about Digg's changes were disappearing. He noticed that his highly dugg negative comments were disappearing from the general view of the site and were only visible if one specifically searched for the "most dugg" comments.[141] One digger quantitatively analyzed the differences between content before and after the version update in order to show that changes must have been made to the voting algorithms to make it easier for the sites that sponsored Digg to get their stories in more prominent positions. After Digg started to roll back its changes and reinstitute the bury button and other lost functions, digger RobertHillberg responded by stating that the site's problems went beyond lost functions, as now Digg was a "dead site filled with crap articles from major media conglomerate sites like time, engadget, mashable, techcrunch, etc. [sic]."[142] After a click-bait slideshow article titled "Famous Logos with Hidden Images" had appeared repeatedly on Digg's front page over the course of two weeks, diggers revolted by posting the Blu-Ray code again and complaining that Digg seemed to have "outsourced its new cheerleading squad." This comment illustrates a xenophobic strain in the Digg Nation that argued for the primacy of American-style democracy and presented the rest of the world as lazy and ignorant. Yet it also showcases how diggers viewed the changes made to their site through a political-economy lens as they lost their stature and other, unknown actors began to take on more of a voice on the site.

This critique was made more damning by the sudden appearance of BP advertisements in 2010, immediately after its Gulf of Mexico oil spill, on the front page that masqueraded as dugg stories. During the BP Gulf oil spill cleanup, pro-BP stories like "Offshore World Looks Good after Gulf Oil Spill, Scientists Say" would appear on Digg's front page and be derided by the user community.[143] While Digg eventually did pull these ads after indicating that its policy was to accept advertisements "from anyone," the presence of this ad furthered the impression that, with version 4.0, the site had changed from one that was primarily concerned with creating a utopian user community to one that was first and foremost interested in garnering corporate and mainstream support. Soon after Digg's version change, Rose stepped down as CEO and the company laid off 37 percent of its staff.[144] It never regained the same level of popularity that it once had and was eventually sold off in pieces to various companies.

In July 2012 the owners of Digg finally achieved their goal of selling the company for millions in profits. LinkedIn bought fifteen of their patents for $4 million (including the "click a button to vote up a story" patent); the *Washington Post* acquired most of Digg's staff for $12 million; and Betaworks bought the actual website for $500,000.[145] While $16.5 million was not a small amount of money, it did pale in comparison to the $200 million for which Google was reportedly interested in purchasing Digg only four years earlier. Viewing Digg's notion of anarchic "digital democracy" as a failure, Betaworks transformed the site by getting rid of all social and voting functionality (though it has recently revived some of these features and connected them to Facebook). Seemingly taking the advice of J. D. Rucker, the company, under its new owner, hired actual humans to find and curate articles on the site—on a genuinely diverse set of topics. Ironically, by casting aside Digg's aspirations for creating a digital democracy, Digg's new owners created a site that was both more equitable for its workers and more diverse in content.

Together, these three examples illustrate how algorithmic recommendations are created within a complex cultural, political, and technological network. Through these recommendations, users come into contact with alternative ways of imagining themselves and consider how their authority, agency, and self-image are always negotiated by these technologies. While these networks tend to adopt and illustrate dominant (and dominating) values, rationales, and logics and present them as perfectly rational rather than always disputed, these examples also illustrate that users often react to them in surprising ways. While the very notion of the "recommendation" gestures toward unequal distributions of power and knowledge, when algorithms pull us, we can push back. The three case studies in this chapter illustrate the different reactions users have had to the many recommendation systems that surround us, including various forms of anxiety, uncertainty, and subversion. Recommendations and the logics they follow have tremendous ideological force and reach, but recognizing this does not mean we must simply give up and cede all authority and agency to them. By assuming that users may sheepishly just take whatever is recommended to them or trust in an algorithm's authority no matter what its output, we risk misunderstanding our relationship to the technologies around us and the power we have to reshape algorithms and the corporations that wield them.

4

Love's Labor's Logged

• •

The Weird Science of
Matchmaking Systems and
Its Parodies

> Well, I filled out my form and I sent it along,
> Never hoping I'd get anything like this.
> But now when I see her,
> Whenever I see her,
> I want to give her one great big I.B.M. kiss.
> She's my I.B.M. baby, the ideal lady,
> She's my I.B.M. baby.
> From the first time I met her I couldn't forget her,
> She's my I.B.M. baby.
> Well we've dated sometime,
> Things are going just fine, and I'd like to settle down with her.
> Just like birds of a feather
> We put 2 and 2 together, and we became one with an I.B.M. affair.
> She's my I.B.M. baby, I don't mean maybe,
> She's my I.B.M. baby.
> —Dave Crump, cofounder of Operation Match

Perhaps no industry is as dependent on recommendation systems as online dating. Many of the most popular and influential services, including eHarmony, PlentyOfFish, OkCupid, and Chemistry.com, rely on recommendations (or what they call matchmaking algorithms) rather than simple searches to bring people together. Until recently, many of these sites would only allow users to contact other users that their algorithms had specifically found for them.[1] Choosing to join such a site instead of one where you can search for people yourself, like Match.com, requires you to place a great deal of faith in their algorithms. Indeed, the belief that matchmaking algorithms "know you better than you know yourself" is omnipresent in the online dating industry. Much of the marketing for matchmaking sites—from OkCupid's lab beaker logo to eHarmony's, Chemistry.com's, and PerfectMatch.com's practice of hiring scientists and academics to vouch for their algorithms—encourages users to believe that their sites are smarter, more trustworthy, and more powerful than the users themselves could ever be. Many of the most popular dating sites present their algorithms as based on scientific studies, and their recommendations are presented as objectively accurate; they train their users to doubt their own desires and trust instead in the sites' proprietary algorithms.

Amarnath Thombre, an engineer at Match.com, perennially the world's most popular dating site, explains that the site incorporates upwards of 1,500 variables to propose possible matches for users.[2] Thombre argues that the closest analogy to this matchmaking process is "Netflix, which uses a similar process to recommend movies you might like," except that, as Thombre states, "the movie doesn't have to like you back."[3] While Chemistry.com has been called a "TiVo for humans," Sam Yagan, the CEO of OkCupid, stated, "We are the most important search engine on the web, not Google. The search for companionship is more important than the search for song lyrics."[4] While he views the search for a partner as "more important," it is telling that Yagan does not think of it as a fundamentally different operation from the search for a song, as dating websites largely match people in the same way that the iTunes Genius recommends songs to users, and both rely on the same underlying collaborative filtering technologies that Pattie Maes helped engineer in the 1990s.

As with other recommendation industries I have already discussed, online dating emphasizes a tension between making a choice and following a recommendation, between personal agency and the ideological apparatuses that surround us. On the one hand, Eva Illouz has argued that these sites are "organized under the aegis of the liberal ideology of 'choice . . .'" and construct the self not just "as a 'chooser'" but also as a "commodity on public display" to be chosen. She asserts, "No technology I know of has radicalized in such an extreme way the notion of the self as a 'chooser' and the idea that the romantic encounter should be the result of the best possible choice."[5] On the other hand, many dating sites emphasize that we need to trust in them specifically because we cannot trust ourselves to make the right choice. Instead, we need to cede our actual, perhaps undiscovered, desires to their recommendations.

Much like Netflix and Amazon, matchmaking sites argue that with so many choices, recommendations are necessary to find true love. "Information overload," or being inundated by choices, has been viewed as a central problem for online dating since its inception. During its early period in the 1990s, the only option for users of these sites was to "manually construct queries and browse huge amounts of matching user profiles" in the hopes of finding a potential sweetheart on their own.[6] Alison P. Lenton, Barbara Fasolo, and Peter M. Todd argue that just as in other areas of life where recommendation systems have proved useful, having more dating choices does not appear to make users happier; it also tends to cost more money and results in people forgetting the choices they actually made and why they made them.[7] In the context of dating, forgetting why someone caught your eye is especially troubling. Rather than a belief that recommendation systems can really help you find your soul mate, it is the simple desire to avoid information overload by limiting choices (regardless of accuracy) that forms much of the justification behind the use of matchmaking algorithms.

At the same time, matchmaking systems are often quite different from the other collaborative filtering systems I have discussed. Often, matchmaking programs incorporate complex personality tests that divide users up so that they can, at times, be matched with those who are most like them and, at other times, be matched with those who are most "compatible" with (i.e., different from) them. The term *matchmaking* itself aligns these practices with the long history of professional religious matchmakers like Jewish shadchans and Hindu astrologers; the connection I made in the last chapter between big data and astrology is even more obvious in the case of matchmaking, where one's matches often appear to be based on little more than a horoscope. (Thus, it is no surprise that "What's your sign?" continues to be a prominent question on most every personality test.)

While these sites attest to their scientific validity, James Houran, who has served as the chief psychology officer at the True.com dating site, and his coauthors point out that there is often very little evidence for these claims.[8] For instance, eHarmony states that its test is scientifically proven and based on thirty years of clinical experience and a nationwide study, but there is no actual proof that any study was actually done or that the test is in any way scientifically valid. Indeed, like many forms of traditional matchmaking, eHarmony was itself started by an evangelical Christian, and I will discuss how the site both reflects and spreads this religious (and specifically homophobic) perspective. Yet, as Caitlin Dewey of the *Washington Post* argues, recommendations created by "the algorithms, the personality profiles, [and] the '29 dimensions of compatibility'" are what "we think of as uniquely 'online' in online dating."[9]

OkCupid even originally started as TheSpark.com, a site designed for high school and college students that oddly combined personality quizzes with Cliff's Notes–esque literature study guides. OkCupid's quizzes, which now number in the hundreds and are continually being created not just by OkCupid itself but also by its users, are now the main draw of the site.[10] These user-created tests form the backbone of the site's personality test collection and collaborative-filtering-based matchmaking algorithms.

Crucially, OkCupid also uses these tests to advertise itself as communal and democratic, in contrast to matchmaking sites like eHarmony, which it presents as authoritarian. OkCupid describes itself as "more like a community art project than an arranged marriage, a throwback to the good old participatory days of the web. Where nanny state sites like True and eHarmony impose upon members their version of what's essential to a relationship, OkCupid lets the people decide."[11] Like Digg, OkCupid frames collaborative filtering as a tool for the creation of democratic and collaborative relationships. Yet these democratic aspirations are disingenuous, as the site's matchmaking algorithms are hidden away and OkCupid always has the final say concerning what goes on its tests and how its matchmaker works. It often expresses contempt toward the scientists that other sites like eHarmony employ, calling them "a bunch of PhDs who probably aren't good enough to be professors at universities."[12] But even as it mocks other sites, OkCupid largely accedes to the demands of venture capitalism and only implements findings and techniques that can make its site cheaper to run. Like Digg and other free sites that rely on collaborative filtering techniques, OkCupid's democratic rhetoric also obscures how its algorithms gather personal information not just to help match people together but also to create and sell personalized advertisements.

Neither the "scientific" nor the "democratic" matchmaking sites, with their various personality tests, are actually able to demonstrate their effectiveness in helping people find love compared to other methods of dating. Thus, if online dating does not actually work, if it is not an improvement on dating more generally, and if its personality tests do not increase one's odds of finding true love, then why do so many companies still employ them, and why do people still go to them? If recommendation systems on dating sites do not work, then what exactly do they do?

In the last chapter I discussed how the companies that produce and use algorithms present them as magical, objective, omnipotent, and omnipresent and how users have attempted to deconstruct this image. Here, I consider how the online dating industry relies on this perception that algorithms are more intelligent than us and can even know us better than we know ourselves when constructing relationships. Matchmaking sites erroneously present their algorithms as reliable and scientifically valid in order to limit our choice of relationships and present this limitation as a benefit. The logic underlying these algorithms echoes a long history of eugenic efforts to create "better" families and, ultimately, a better species. Following this logic, these matchmaking sites end up promoting conservative relationships that perpetuate gender binaries, and they place value on conformity and sameness over difference in their couplings, which leads to the racial segregation of their users.

To describe how matchmaking algorithms developed to encourage the creation of eugenically enhanced families and otherwise conservative relationships, I begin by tracing the long history of dating practices that have relied on statistics, psychological personality tests, and computers to create successful relationships. In the process, these efforts attempt to enforce a one-size-fits-all approach to relationships and the belief that everyone must experience love in the same way. After tracing the history of computerized dating, I discuss how eHarmony led the modern online dating industry to rely on recommendation algorithms as a marketing ploy that has led to exclusionary, misogynistic, homophobic, and racist practices. Yet I also show that many users have continually critiqued and challenged the way these companies use their algorithms; as I illustrate, algorithmic matchmaking has led to an increasingly combative and often parodic reaction by critics and users alike who employ these programs to their own ends.

While online dating was once thought of as the last resort of the desperate, clicking on and swiping through possible wooers has rapidly become de rigueur. In 2002 *Wired* magazine proposed that "twenty years from now, the idea that someone looking for love won't look online will be silly akin to skipping the card catalog to instead wander the stacks because the right books are found by accident."[13] This article goes on to indicate that while we have a "collective investment in the idea that love is a chance event," online dating is replacing such serendipity with the logic of an efficient "marketplace of love." The pleasures and erotics of wandering a library and coming across something (or someone) you did not even know you desired seem to be lost here; instead, online dating sites are typically represented through the metaphors and logic of consumerism, and the serendipity of browsing becomes the enemy of efficient lovemaking.

If anything, *Wired* was off by at least a decade. By 2008, online dating was nearly a billion-dollar industry in the United States alone, and from 2005 to 2012, nearly 35 percent of Americans met their spouses online.[14] Online dating continues to soar, and from 2013 to 2015, the number of eighteen-to-twenty-four-year-olds on sites tripled and the number of seniors using sites doubled, making online dating a multi-billion-dollar industry. There is now a giant ecology of dating sites catering to virtually any type of human imaginable. There are sites for people with allergies, STDs, criminal records, or even an interest in ghosts (the Supernatural Dating Society). Whether you are a clown (or clown curious), a sea captain, a Zoroastrian, a Jain, or a Satanist, there is a site designed for you. And if you cannot find a dating site for people who share your particular interests or fit your "type," dozens of companies, including Dating Pro, DatingScript, Chameleon Social, AdvanDate, and White Label Dating, will make one for you or provide all the technology (including match-making software) needed to make one yourself.

Yet, as online dating has become quotidian, it has also become a target of increased critique, derision, and controversy concerning how it commodifies us. While *Wired* presented this consumercentric future of romance as rational if not entirely utopic, many people since have instead sneered at such efforts. For instance, Sherry Turkle describes how "technology proposes itself as the architect of our intimacies" but ends up only making us feel more lonely and insular.[15] Perhaps most insightfully, Illouz considers how, at every stage of online dating, sites construct users as standardized products in a rational marketplace dominated by the logic of supply and demand. She argues that as these sites construct us as consumer goods, we begin to think of ourselves as such as we "look for ways to improve [our] position in the market."[16] At the same time, these sites train us to increasingly refine our tastes and manage ourselves by also treating potential lovers as commodities. Even in articles that neutrally describe the workings of online dating, or actively propose new technologies for these sites, many cannot resist using puns and wordplay that critique (intentionally or not) the marketization of romance. For instance, David Sköld calls dating sites "marketplaces" and considers how they offer a "commercial construction of a perfect date"; Lenton, Fasolo, and Todd refer to these services as "'shopping' for a mate"; and Jeana H. Frost and her coauthors unironically refer to their users as "experience goods."[17] To various degrees, all of these authors argue that this hypercommodification and rationalization of dating too often leads to superficial (if not outright inhumane) relationships wherein the desire to continually trade up for ever better matches is fostered.

In the same vein, online dating also facilitates various forms of racism. If matchmaking systems are supposed to highlight people we may have (for whatever reason) skipped over or not considered, one might realistically hope that this would also lead to them encouraging people to date across racial, ethnic, national, economic, educational, and ideological lines that are perhaps less divisive than they may at first appear. Unfortunately, this is far from the case. Instead, most sites feature matchmaking algorithms that allow users to choose (among other things, like class and salary level) which races they are willing to date members from and, with one click, exclude those who do not fit this list. This practice, which allows people to exclude others from consideration based solely on their race, is a modern and all-too-common form of accepted segregation. Many sites like eHarmony, Chemistry.com, and PerfectMatch.com follow a deeply problematic "separate but equal" logic by directing minorities and members of other "niche" groups toward partner sites that specialize in these particular communities, whether they be Latinos, African Americans, homosexuals, Christians, Jews, Muslims, or members of another group. As one *Time* article notes, these dating sites are one of the last vestiges of real racism and allow users to never even be presented with the possibility of meeting someone of a different background if they do not think they could be in a relationship with the person.[18]

These other niche sites do not typically feature matchmaking systems or personality tests and instead assume that these users want to search for partners themselves. This assumption causes matchmaking systems to become a tool primarily of mainstream users (not part of any niche community) who are largely coded as white and bourgeois. Just as Neiman Marcus used recommendation systems to market its wares to rich customers surrounded by an abundance of commodities, dating sites use the same logic to connect similar users surrounded by an abundance of profiles. While the rich, bourgeois audiences that sites like OkCupid and eHarmony attract become associated with this abundance of admirers (and the concomitant need to cull these users down to a manageable handful), the proletariat and minorities become instead associated with scarcity and the expectation that there would be so few people interested in dating them that matchmaking would not be necessary. Thus, recommendation systems become tools of structural inequity.

While specifics on how racial preference works are not typically made public, OkCupid did publicize statistics showing that its users (and those of other sites OkCupid owns) in 2014 were much more likely to rate others of their own race higher than others, with the exception of black men who expressed a slight preference for Latina and Asian women over black women. Its data showed many troubling findings, including that black women and East Indian men were least likely to get responses from others on the site and that from 2009 to 2014, its user base had actually become more racially biased.[19]

Rather than indicate ways that its site or algorithms may have contributed to this racial bias, OkCupid argued that its users came to its site already with racialized preconceptions that it was in no way responsible for shaping or undoing. At the same time, it specifically avoided calling its users racist by arguing both that they are "not any more [racist] than anyone else" (i.e., the "everyone is racist" defense) and that people "cannot really control who turns them on—and almost everyone has a 'type,' one way or another."[20] On the one hand, OkCupid does appear concerned about racism as a "cultural trend," but on the other, the company is averse to recommending people change their dating habits or even acknowledging that their individual actions are racist. While I agree that getting rid of the racial preference options on dating sites would not magically eliminate racism on these sites (as people can be endlessly creative when it comes to being awful to each other), that in no way justifies their creators' decision to make it so easy to be racist on them. These sites justify these blanket racial filters by arguing that people cannot help who they are attracted to and therefore that encouraging people to date beyond their comfort zone is pointless and unprofitable. Yet, as historical transformations in what is considered beautiful indicate, attraction is a learned behavior shaped by culture; any claims that attraction is based in biology (especially in relation to racial attraction) are extremely overblown and only serve to justify racist preferences.

To top it all off, there is no evidence that online dating works at all. Even as more and more people use these sites and find relationships on them, several studies have shown that the chances of finding love are next to random. Psychologists and sociologists have persuasively argued that while online dating sites may make it easier to access a list of potential partners, this benefit is undermined by the ways dating sites present information about users and the faulty logic they use to recommend particular matches over others. For instance, psychologist Eli J. Finkel and his coauthors argue that while online dating offers the possibility of meeting many more people than you might otherwise, people are actually quite bad at getting a sense of who they might be compatible with from user profiles (the information and images that users provide for others to browse through); people must first meet someone face to face before they can tell, but if they do not like the other person's profile, they may never get that chance: "Online dating profiles reduce three-dimensional people to two-dimensional displays of information, and these displays fail to capture those experiential aspects of social interaction that are essential to evaluating one's compatibility with potential partners."[21] Furthermore, "the ready access to a large pool of potential partners can elicit an evaluative, assessment-oriented mindset that leads online daters to objectify potential partners and might even undermine their willingness to commit to one of them. It can also cause people to make lazy, ill-advised decisions when selecting among the large array of potential partners."[22] As Lenton, Fasolo, and Todd's study attests to, many online dating sites try to alleviate these problems with recommendation systems, which limit the number of matches users see to those they may be more compatible with.[23] While Finkel and Susan Sprecher agree that this is a perfectly logical solution, they also argue that, unfortunately, "the mathematical algorithms at matching sites are negligibly better than matching people at random."[24] Furthermore, they argue that while online dating sites typically base their matches on either the premise that similar people get along together better or the premise that opposites attract, there is no evidence indicating that either principle has any impact on the quality of relationships.

These findings are largely echoed by several other studies in academic journals that have shown that online dating leads to fewer committed relationships and that matchmaking algorithms are ineffective.[25] While there are several studies that stick up for online dating by proposing that "more couples meet online than at schools, bars or parties" and that simply having internet access boosts marriage rates, these studies do not evaluate the happiness of couples or even whether they met on a dating site at all. The lion's share of studies that argue online dating works are published by people working at the dating sites themselves and rely on proprietary data that cannot be evaluated by others. At best, as the *Washington Post* reported, dating sites are no worse (though probably no better) than a neighborhood bar.[26] While I am not convinced that online dating is necessarily any more awful than any other way of meeting people, it does appear that by relying on a consumerist logic and recommendation technologies, online dating has managed to automate the vilest parts of contemporary dating culture.

With all of these systemic problems, one may be tempted to ask both why dating sites still employ matchmaking systems and why people use them at all. In earlier chapters, I have left the assumption that recommendation systems are designed primarily to guide you toward things or information you might actually want largely unchallenged. In an industry like online dating, where pleasing a customer may mean that you lose them forever, this assumption is much more specious. Brian Bowman, a former Match.com executive, stated, "Everyone knows that all personality profiling is bullshit. . . . As a marketing hook, it works great."[27] While matchmaking systems may not accomplish what they are ostensibly supposed to, that does not mean they do nothing at all. Indeed, they do facilitate the creation of certain kinds of relationships over others; each algorithm encourages a different kind of relationship, and dating sites use these differences to, as Bowman put it, declare the "unique value" of their sites to the niche audiences they are trying to attract. By tracing how contemporary online dating and personality tests came to be, I will illustrate how economics and culture have shaped and continue to shape these algorithmic practices, which in turn structure many of our most intimate choices and values.

Personality Tests and Predicting Relationships

Modern online dating has its roots in the long history of using psychometric, statistical, and computer sciences to engineer "better" couples, families, children, and societies. Many of the personality tests and matchmaking algorithms now used by dating sites rely on techniques developed out of Sir Francis Galton's nineteenth-century founding theories on psychometrics, eugenics, and the lexical hypothesis of personality. A half cousin of Charles Darwin, Galton was the first to use questionnaires and apply statistical methods to the study of human behavior and intelligence. Convinced that everything is measurable, Galton attempted to statistically calculate and judge everything from beauty to boredom.[28] He planned to use these data to figure out which people were "best" in the hopes of getting them to have more children and thereby improve the human species. Galton found that richer families tended to score higher on his tests, and, instead of considering the role that socioeconomic injustices and prejudice played in this process, he argued they must be rich and successful because of their better genes. Learning entirely the wrong lesson from his findings, he thought that the state should pay high-ranking families and other eminent people to intermarry at an early age so that they might have more babies and eugenically improve the human race.[29]

Along with beauty, intelligence, and many other facets of humanity, Galton created tests to measure personalities. While it does not seem as though one would necessarily need a test to identify one's personality, as Sigmund Freud would later argue, many of our motivations, actions, and desires are beyond our conscious awareness, and we do not actually know ourselves very well. This insight implies that personality tests may help us understand ourselves and our actions better; it is also the basis for the modern belief that recommendation algorithms may know us more accurately than we know ourselves.

Galton hypothesized that personality differences between people would be encoded in language and that, therefore, one could create a comprehensive list of personal attributes by focusing on the words used to describe such personalities. In 1936 Gordon Allport and S. Odbert applied this lexical hypothesis to the creation of a list of more than 4,500 adjectives that described human traits. In the 1940s Raymond Cattell culled this list down to 171 unique descriptors, which he then arranged into sixteen discrete personality categories.[30] In 1961 Ernest Tupes and Raymond Christal reorganized these personalities into five broad categories. Since then, several psychologists have redrawn these categories in various ways, and in 1981, psychologist Lewis Goldberg reformulated them as follows: "Extraversion or Surgency (talkative, assertive, energetic)"; "Agreeableness (good-natured, cooperative, trustful)"; "Conscientiousness (orderly, responsible, dependable)"; "Emotional Stability versus Neuroticism (calm, not neurotic, not easily upset)"; and "Culture (intellectual, polished, independent-minded)."[31] While there continues to be debate concerning the exact names and definitions of these categories (not to mention how the Culture category naturalizes the conflation of independence with wealth), psychologists now widely use these "Big Five" personality traits to describe human personality at an extremely broad ("big") level. Psychologist Dan McAdams has referred to these traits as extremely vague and general: "It is the kind of information that strangers quickly glean from one another as they size one another up and anticipate future interactions."[32] Psychologists and sociologists use this set of traits and the tests used to define them largely to study how different backgrounds and stimuli affect personalities.

Like the Big Five traits, the Myers-Briggs Type Indicator (MBTI) is also often used to hypothesize relationship satisfaction via online dating sites.[33] The MBTI is a personality test similar to the Big Five except that it generates four personality categories instead of five and these categories are based entirely on

Carl Jung's typological theories.[34] It was originally developed in 1942 to help women who had never worked before identify the kind of job for the war efforts that they would be most "comfortable and effective" at doing.[35] Many industries now employ it when making hiring decisions, and the recession that began in 2007 has only increased its use, as "many firms want to hedge their bets and cannot afford to pick the wrong people. Tighter profit margins also mean working under more stress and companies want to make sure their employees get on. Disagreements are costly and inefficient."[36] As quoted in *The Cult of Personality*, Annie Murphy Paul criticizes these tests and the desire to break down identity into manageable and commodifiable "neat little boxes" to serve our purposes as "quintessentially corporate America."[37] Using personality tests like the MBTI to guide decisions in all areas of life, we end up constructing all our relationships as labor, using the logic of productivity and efficiency on which it depends.

In her explanation of the history and function of the MBTI, Isabel Briggs Myers explains that while people with opposing personality profiles may work well together, relationships can be more harmonious if the people in them have similar if not identical personalities. People with opposing personalities can have great relationships but must always "take the necessary pains to understand, appreciate, and respect each other."[38] Those with similar personalities theoretically share the same perspective on the world and can more easily understand and appreciate each other's reactions and feelings; in other words, the MBTI values similarity because it ostensibly leads to a more efficient and productive form of relationship. In the process, this focus on similarity reinforces practices of segregation by explicitly dissuading users from interacting with those whose personality types appear to diverge from their own.

Yet, while the MBTI values similarity in relationships, many online dating websites now use modified versions of the MBTI to match people who are not similar but rather complementary or opposites. Chemistry.com, True.com, PerfectMatch.com, and Yahoo! Personals all incorporate personality tests loosely based on the MBTI, and all focus on complementarity in finding people matches. As sociologist Pepper Schwartz, the creator of and spokesperson for PerfectMatch.com's Duet Total Compatibility System, explained, "We all know, not just in our heart of hearts, but in our experience, that sometimes we're attracted, indeed get along better with somebody different from us."[39] While Schwartz appealed to common sense, there is little evidence for this assertion. Studies have shown that certain personality similarities (especially agreeableness) correlate with satisfying relationships, but the evidence is largely contradictory and shows only that some people have found happiness with people that are similar to them and other people prefer people who are different from them.[40]

While catchy, in practice, this "opposites attract" logic has led to a problematic gendered stereotyping. For instance, Chemistry.com's test tries to figure out how feminine or masculine you are, and its matchmaking algorithms connect you with people on the opposite side of the spectrum. For instance, if your answers indicate that you have stereotypically extremely masculine traits (including spatial acuity and a focus on rationality), the test will match you with people who have extremely feminine traits (including emotional acuity and empathy). This test reinforces a harmful gendered binary that opposes feelings with thoughts and individualism with community; such logic creates relationships wherein one person is always subordinated as the supporter of the other, who is framed as the leader or maverick.

Chemistry.com's test was created by Helen Fisher, a biological anthropologist who studies how hormones and other biological factors affect our personalities and who we are attracted to. She argues that the way we each think and behave is a product of evolutionary biology rather than of our environment and upbringing. Much like the Big Five personality traits or the MBTI, Fisher breaks personality into four styles: creative explorers, sensible builders, reasoning directors, and intuitive negotiators. While everyone illustrates aspects of all of these styles, she argues that most people are dominated by one or two styles, which are determined by our hormone balance. Much like an ancient doctor interested in healing a patient by rebalancing his or her humors, she associates these different personality styles with the balance of dopamine, serotonin, testosterone, and estrogen in our bodies. She associated dopamine with creativity, serotonin with sensibility, testosterone with rationality, and estrogen with intuition. Since women typically have more estrogen and men more testosterone, Fisher naturalizes a long-standing sexist stereotype that men are naturally more rational and women more intuitive.

Fisher's theory structures Chemistry.com's personality test and matchmaking algorithm. Many of the questions attempt to measure users' hormone levels in order to define their personalities. For instance, the first question asks users to measure their ring and index fingers, as research indicates that those with longer ring fingers may have been prenatally exposed to testosterone. In turn, this test takes the extra jump of assuming that being prenatally exposed to testosterone will necessarily affect one's current level and decide how rational the individual is capable of being. Other questions ask users to judge the relative size of various objects and lines, while another asks whether a variety of people in photos have real or forced smiles. These questions assume that having an aptitude for spatial relations and geometry is a masculine trait and being able to identify emotions is feminine. This stereotyping is compounded by the gendered occupation titles that Fisher and Chemistry.com give to those who exhibit personalities associated with each hormone: those with rational, testosterone-fueled personalities are called "directors"; those with supposedly high levels of estrogen and intuitive personalities are called "negotiators"; creative types filled with dopamine are termed "explorers"; and those ostensibly made stable and caring by an increased level of serotonin are named "builders."[41] The description of each personality type is also laced with sexist assumptions, including that directors are goal oriented, logical, individualistic, competitive, and prone to explosive anger.[42] In contrast, the negotiator type is imaginative, intuitive, and empathetic. Following a long-standing feminine stereotype, negotiators, supposedly due to their high estrogen levels, are caring, careful, and good communicators, though they may talk too much and appear indirect. These definitions naturalize harmful stereotypes that make masculinity a position of dominance and femininity one of support.

After completing the personality quiz, users receive their "personality score" and an explanation of what their answers say about them and what kind of person they might get along with. While the site states that "you are made up of four personality types," it tells you the two that "have the most influence over your personality."[43] Everyone is labeled as either a director or a negotiator (i.e., feminine or masculine) and a builder or an explorer (inhibited or impulsive). In effect, everyone, regardless of his or her actual gender identity, is defined by how stereotypically feminine or masculine he or she appears to be.

Like PerfectMatch.com and Chemistry.com, OkCupid's personality test also relies heavily on the MBTI to engineer potential relationships. However, rather than using MBTI-specific categories, OkCupid playfully classifies people based on whether they are meticulous or random, brutal or gentle, a master or a dreamer, and interested more in love or sex. Based on this test, if you are a woman, it is possible to be labeled as a peach, a vapor trail, the dirty little secret, or Genghis Kunt, among others.[44] The sexism here is obvious.

While complementarity can mean many things, in practice, on Chemistry .com, it largely means that director types are matched with negotiator types and builders are matched with explorers. The more feminine your answers make you appear, the more masculine your matches will be. I, a director/explorer according to its test, have never been matched with anyone other than negotiators, or those who had not as of yet taken the test (possibly because they had not made it to that early point in the subscription process or because IAC [InterActiveCorp, the owner of Chemistry.com] may match people across its many dating sites). In turn, if the test decides you are a very feminine man, it will match you with very masculine women. The same logic applies if you are looking for someone of your own sex, as the site will still try to match you with someone who registers as having opposing gender attributes.

While this logic breaks a connection between gender and sex, it does so by reifying the gender binary and the belief that male attributes inherently pair well with feminine ones. Studies that have supported the idea of complementarity have long been used to justify the idea that women are essentially different from men and that the woman's natural place is in the home while the man should be the breadwinner. These discourses support what Diane Negra has labeled larger postfeminist strategies of essentializing women around "a common set of innate desires, fears and concerns" related to "heterosexual partnerships and mothering."[45] As Anthea Taylor has discussed, while today this emphasis on complementarity promotes "partnership between equal individuals," this belies a contemporary reality in which complementary partners in relationships, families, or marriage often experience a power asymmetry, often around work and home.[46] Discourses about complementarity have historically stressed differences between sexes, with women being tied to nature and the home and men being part of the world of culture and the office. These narratives indicate that as women enter the workplace and the public sphere, they lose their ability to take care of the private sphere, which Taylor argues is too often seen as the only "legitimate affective investment for women."[47]

Online dating personality tests often support this position by matching these "masculine" women with "feminine" men who have themselves entered the private sphere. The gendered discourses on these dating sites encourage one person to take on masculine characteristics of power and control while the other takes on feminine characteristics of cooperation and reconciliation. The difference in this postfeminist situation is that this logic of complementarity versus similarity is applied to all parts of life, and these dating sites no longer strictly expect that the man will take on the masculine characteristics and the woman the feminine ones. These values continue to be gendered, but it is now acceptable on these sites for the man to be gendered female and the woman to be gendered male, as long as they together are complementary. What is most important on these matchmaking sites is that the gendered power dynamics of heteronormativity are preserved.

Computers and Dating

Matchmaking companies very early on began using computers to calculate the results of personality tests to help people find true love (or, more often, just a date). While one may be forgiven for thinking that matchmaking systems began with eHarmony, they have been around nearly as long as computers themselves. Throughout this long history, the central tension surrounding the industry has been whether online dating should or can be a tool for creating monogamous relationships or whether it is most capable of providing casual dates. In contrast to companies dedicated to providing casual dates or paid escorts, since the industry's earliest moments, dating companies used personality tests and matchmaking programs to advertise themselves as scientific tools for the creation of happy marriages. As with more recent sites like Chemistry.com and PerfectMatch.com, in order to convince potential customers that they could create relationships at all, early matchmaking sites argued that they knew what a good relationship consisted of and that their algorithms contained the mathematical formula for true love and happiness.

Many of the original tests and techniques for computer dating were developed in the 1940s in response to changes in gender norms and urban lifestyles during World War II. Eleanor Roosevelt proclaimed that the country's third-greatest sociological problem was that strangers in a big city have difficulty in "meeting suitable companions of the opposite sex."[48] In response, Herbert Gersten, a Rutgers student who heard a related talk by Vassar professor Joseph Kirk Folsom on the "solitary and/or mismated state of many young metropolitans," started an early statistical dating service called Introduction, a Service for Sociability, in Newark, New Jersey, in 1942.[49] This service was one of many introduction services and matrimonial bureaus around the country that opened to take advantage of this growing market of urban daters.[50] Gersten convinced several "civic bigwigs, clergymen, educators and clubwomen" to fund his "idea of harnessing Cupid to statistics" and proceeded to experimentally match five hundred random people up in dates in order to perfect a statistical rating system.[51]

Introduction asked applicants "to fill out blanks giving their age, height, religion, education, occupation, interest and hobbies," along with recommendations from family doctors or ministers. For a two-dollar registration fee and twenty-five cents per date, it paired people off if they had similar hobbies, education, heights, and preferences for what to do on a date. A *Life* magazine article and advertisement for Introduction showcased how anyone from any class and background could profit from this scientific form of dating. In order to emphasize this point, Gersten discussed the potential of using scientific dating services to bring together vegetarians, a small group "who will consort only with non meat-eaters like themselves."[52] It also featured a photomontage of two college students meeting through the service and their first idyllic date to a dance and carnival. While based in Galtonian psychometrics, these dates largely confirmed stereotypes about attraction, including that women prefer "tall men with money" and "men are content with a good sport."[53] Importantly, Introduction also advertised itself as a licensed real estate broker, ready to make money by selling houses to the happy couples it was busily creating. While never more than a local novelty business with a few hundred clients, Introduction was one in a long line of dating services interested in applying statistics and personality tests to make dating both more logical and more profitable.

In 1957 George Crane started the Scientific Marriage Foundation, one of the (if not the) first computer dating organizations. Crane had long been a psychologist famous for his widely syndicated newspaper column and many pamphlets that featured tests in which married couples could rate their relationships. One of his more famous 1930s pamphlets, *Tests for Husbands and Wives*, features a long and varied list of questions created through "the composite opinions of

600 husbands [and 600 wives] who were asked to list the chief merits and demerits of their wives [and husbands]."[54] Questions for husbands included such things as whether they call "'Where is . . . ?' without first hunting the object" and whether they see "that wife has orgasm in marital congress." For wives, questions ranged from whether they squeeze "tooth paste at the top" to whether they are "willing to get a job to help support the family."[55] As can be seen, while some of the questions appear to promote traditional gender roles, others support the importance of equality between partners.

Crane marketed his early theories and personality tests through the Scientific Marriage Foundation. Via this service, customers would first complete an interview with a counselor and send in references and photographs to Mellott, Indiana, where an IBM sorting machine would pair couples off according to their stated preferences and desires. Much like eHarmony, Crane's company relied on a combination of "scientific" surveys and the input of various religious leaders to create matches.[56] While the company only lasted three years, it advertised that during that short period it was able to create five thousand marriages.

Though the Scientific Marriage Foundation failed to take hold, hundreds more companies arose throughout the 1960s to cater to changing dating demographics and a perceived lack of satisfactory partners.[57] Computerized matchmaking became an important part of a mushrooming "institutionalized dating" industry of singles bars, hotels, apartments, washaterias, and so on in the 1960s, created to serve "America's eight million singles and tap the $50 billion they spend each year—mainly in pursuit of each other."[58] Likewise, as Marie Hicks has explored, while much of this industry focused on casual dating, computerized matchmakers advertised instead to those in search of long-term commitments and marriage.[59] A 1967 *Life* magazine article on new dating trends echoed Eleanor Roosevelt's earlier worries and argued that during this feminist period when sexual mores and gendered dating rules were loosening, "free as a lark" singles complain "it has become almost impossible to meet members of the opposite sex, especially in the cities. Life, says one secretary, is an endless subway ride, a day at the office, an evening at home in a double-locked apartment."[60] This article asserted that "big business" in the guise of computer matchmakers serves their demographic of young, single people by offering to manage and minimize the time constraints and anxieties of feminism and the sexual revolution through their focus on creating lifelong monogamy rather than an opportunity for sexual experimentation. While singles could meet a "leftover" date from a friend or family member, computer matchmakers offered singles for $5–$150 depending on the service, a safe way to meet interesting and compatible people in an increasingly "mobile" world.[61]

Just as yentas have been imagined as being better at creating successful matches because of their lifetime of experience, advertisements have often touted computers' great matchmaking capabilities because they can take into account more variables and compute more data than any human. One advertisement for the computer dating company Compatibility Research from 1969 promoted the computational power of their "IBM 360/40 Computers," which "will do more in an hour than a highly qualified individual can do in a year."[62] Companies like Compatibility argued that there was room in the marketplace for their particular service by asserting that while a short-term fling in the middle of the sexual revolution was easy to engineer, lifelong romances were now beyond the capabilities of most humans to create.

Many of these services tried to compare themselves favorably to newspaper personal ads and other methods of finding partners by asserting that their matchmaking software made dating safer and could create long-lasting relationships, while personal ads could be very dangerous and only led to one-night stands. As Harry Cocks has argued, personal ads in the 1960s became connected to homosexual and countercultures, and the police often censored ads for prostitution, swinging, and other nonheterosexual, nonmonogamous relations.[63] These views were strengthened in 1968 when *Exit* and *Way Out* magazines in England were found guilty of conspiring to corrupt public morals through their personal ads.[64] Yet, as Hicks has argued, the banning of newspaper ads for dating services encouraged dating companies to create services geared more toward conservative, long-lasting relationships to set them apart, but their fringe status also made them attractive to younger customers. For instance, Com-Pat, an early British dating service originally started for women who needed male companions to attend events, advertised its services on illegal "pop pirate" radio stations because they were the only ones who would allow it. While it largely tried to create conservative relationships, its advertising made it appear "cool and cutting edge."[65]

Other companies, such as Dateline, which from 1967 to the mid-1990s was the most profitable dating service in Britain, appeared to be more "indiscriminate" and "unscrupulous" concerning how it matched users together and catered to users more interested in finding casual dates than long-term relationships.[66] Dateline was criticized both for the amount of delicate personal information it requested from its users, including their views on premarital sex and communism, and for apparently not using this information to create good matches. In response to accusations that Dateline's algorithms did not match people together in any logical way, Dateline's founder, John Patterson only retorted, "Even if his members have nothing else in common they have at least all joined Dateline."[67] Given Bowman's comments concerning how personality profiles are "bullshit" that is only good for marketing, Patterson's reasoning seems to hold sway today.

During the late 1960s and 1970s, the question of whether dating companies were providing monogamous relationships or one-night stands tapped into larger fears during the sexual revolution over whether these technologies were transforming the definition of modern relationships or were helping to preserve older models of heteronormative monogamy. At the same time, while computerized dating companies largely focused on creating monogamous, heterosexual relationships in a safe environment, a handful of companies focused on a homosexual clientele. John Hendel at the *Atlantic* attributes this growing demand for computer dating to the sexual revolution and the related expansion in "our society's moral flexibility, technology, and the enthusiasm of young date-hungry entrepreneurs, particularly during the second half of the 1960s."[68] While many new companies catered specifically to heterosexual customers interested in ever-lasting love, some, like Man-to-Man, were designed specifically for homosexual men who wanted to "forget standing on street corners—being harassed by the authorities—searching through smoky bars—Now! Do it—the easy-scientific way."[69] As Lucas Hilderbrand has discussed, an ad for this service promised to "rationally make romantic matches" via "scientific" punch cards in order to make "romance efficient and clean" and eliminate the "thrill of transgression."[70] While cruising had been framed only as a way to pick up a trick, the ability to "cruise by gay computer" promised the possibility of a long-term, homosexual relationship.

This logic that privileges the extensive knowledge and efficiency of yenta-like technologies tends to support "traditional" relationships that too often benefit husbands over their wives; they compute only the most conservative of logics. Hicks points out that technologies have historically been used not to create new forms of relationships but rather to conserve older forms during periods when they are under threat.[71] One of the most influential of these companies was Project TACT (Technical Automated Compatibility Testing), which operated on New York's Upper East Side starting in 1965. Clients paid five dollars to enroll in Project TACT, a computerized dating service that included a hundred-question personality test.[72] Nick Paumgarten has noted that, as in many of these tests, the questions were often highly gendered: while men were "asked to rank drawings of women's hair," women were asked "where they'd prefer to find their ideal man: in a camp chopping wood, in a studio painting a canvas, or in a garage working a pillar drill."[73] Questions like these reinforce the idea that men are defined by what they do, while women are defined by how they look. Clients' answers were then fed through an IBM 1400 Series computer, which would "spit out your matches: five blue cards, if you were a woman, or five pink ones, if you were a man."[74]

During the mid- to late 1960s, computerized dating programs like this were viewed as a response to the positive gains of feminism. The Upper East Side was known as a "an early sexual-revolution testing ground" full of "young educated women who suddenly found themselves free of family, opprobrium, and, thanks to birth control, the problem of sexual consequence."[75] Patricia Lahrmer, the first female reporter at New York radio station 1010 WINS, focused her first feature story on TACT and on how modern New York couples meet. In her story, she compared TACT to Maxwell's Plum, one of the first singles bars in New York, stating that they were both places where "so-called 'respectable' single women could patronize on their own."[76] While her story celebrated how the sexual revolution had led to more women being able to search for dates and ask men out, it also separated these positive effects from sexual experimentation, viewed here as disreputable. Stories like hers promoted computerized dating as being clean and safe in order to ease women's fears of potentially being tagged as sexually promiscuous if they used the service. Full of contradictions, these programs promoted traditional relationships in ways that might seem counter to feminism but were also ostensibly viewed by women as a better way to date.

After the success of projects like TACT, many similar programs were started around the country. In response to these contemporary trends regarding changing gender norms, three Harvard students envisioned Compatibility as an answer to the "irrationality of two particular social evils: the blind date and the mixer."[77] These students used the sexual revolution as a justification for their service and envisioned computers as a rational response to the difficulties of contemporary dating practices, which were themselves responses to changing gender expectations and a rapid opening up of educational possibilities to women as a result of second-wave feminism.

Their first effort, in 1968, called Operation Match, was aimed at using computers to generate dates and matches for college students that were "more permanent than a mixer, and more fun than a marriage bureau."[78] Even though Operation Match asserted it led to long-term relationships, in contrast to mixers, it actually became popular with people who specifically wanted to "keep dating casual."[79] With help from friends in the Harvard Sociology Department, Operation Match's creators formulated their "scientific" questionnaire in two weeks. Soon after, a member of the company, Vaughan Morrill, appeared on *To Tell the Truth*, a CBS quiz show, to promote the company and computer dating. On the show, Operation Match was lauded for being able to "take the blindness out of your next blind date," but it was also criticized for demystifying romance and rationalizing love.[80]

Following Morrill's CBS appearance, Operation Match sponsored the trip of Vicki Albright, a University of California, Los Angeles, student (who had just been selected by the UCLA Law School as their "Woman of the Year"), to Harvard, where she was set up by its high-speed Avco 1790 computer with a date with a Harvard man. Stories of their date and Operation Match circulated across the country and resulted in Compatibility Research's opening nine offices across the United States and an enlargement of its customer base to also include high school students, who, for around twenty-five to fifty cents, could be matched with other students for school-sponsored social events.[81] It also opened several international offices, and a London television advertisement showcased the "transistorized pairings, hundreds of them, dates by the dozen and many marriages" created by Operation Match.[82] At the height of its success, it reportedly had over a million customers, who each had paid three dollars to sign up.

In the 1980s computerized matchmakers gradually gave way to online matchmaking services that were often just postings on bulletin board sites. However, in 1986, Matchmaker.com became the first modern dating site to offer a multiple-choice personality test, essay questions, and a recommendation system. Its test asked users a wide range of questions, including what kind of temper they had, what languages they spoke, what kind of relationship they wanted, how they dealt with conflicts, and what their feelings on conformity were.[83] The program would then generate a list of possible matches on the site for the user and list them in order of who had given the most similar answers. Throughout much of the 1990s, Matchmaker.com was the most successful dating site, with four million members, but it dropped in popularity in the 2000s due in part to competition and a lack of advertising.[84] With its test and rating system, however, Matchmaker.com's influence pervades.

eHarmony and the Modern Matchmaking Service

Companies like Operation Match, Project TACT, and Matchmaker.com are direct precursors to contemporary online dating services like eHarmony. Like these earlier companies, online dating sites use the same rhetoric about the changing nature of society and the increasing difficulty of finding and sustaining love to argue that their personality tests and matchmaking software are necessary for creating stable, ever-lasting relationships in an otherwise chaotic universe. They advertise that they can supply safe and empowering dates if and only if users give up much of their control over the kind of person these dates are with (and, in some cases, what the date consists of). For example, until 2016 eHarmony "guided communication" between users by selecting the questions they could ask each other and the topics they could discuss before they could actually meet in person.[85] Like these earlier dating companies, modern dating sites now advertise the empirical, scientific nature of their personality tests and assert that central life choices are too difficult for individuals to make on their own.

eHarmony was one of the first, and by far the most influential, contemporary dating sites to employ personality tests and matchmaking algorithms. Most other matchmaking sites, including Chemistry.com, OkCupid, and PerfectMatch.com, were created specifically in eHarmony's image and as its competition.[86] eHarmony was launched in 2000 by founder Neil Clark Warren, an evangelical marriage counselor with a master of divinity degree, and Galen Buckwalter, a psychology professor at the University of Southern California. Warren, a frequent guest on Pat Robertson's *500 Club* conservative talk show, also authored several self-help guidebooks, including *Finding the Love of Your Life: Ten Principles for Choosing the Right Marriage Partner* (1994) and *Finding Contentment: When Momentary Happiness Just Isn't Enough* (1997). Many of his books focus on the difference between long-term and momentary happiness and describe the importance of forming a relationship around what he describes as essential and core aspects of oneself rather than fleeting feelings.

At the time (and still today), Match.com dominated the online dating industry with forty-five million users.[87] Rather than assume that people are led by unconscious desires that they cannot fully understand, Match.com starts from the premise that people do know what they want in a relationship and should be able to search for it themselves. It features a relatively simple interface with no personality tests or recommendation algorithms. Instead, it offers many filters and asks users to search for potential dates on their own and tinker with their search parameters until they find what they want.

While Match.com focuses on making it easier for users to search for others on their own, eHarmony argues that people do not really know what will make them happy over the long term and therefore cannot be trusted to search for it on their own. Similar to the argument presented in Warren's books, eHarmony's mantra is that we cannot help but be guided by fleeting feelings and therefore need an "objective" outsider perspective (in this case, an algorithm) to allow us to consider only those aspects of ourselves that Warren considers essential (or rational in a Cartesian sense). His collaboration with Buckwalter illustrates eHarmony's attempt to appeal to those interested in using religious principles and science to locate love and create a family. eHarmony echoes Galton's eugenic sensibility through its focus on compatibility and similarity in matchmaking in the hopes of creating successful families. Like Galton, eHarmony also has the extremely broad goal of lowering national divorce rates and improving relationship happiness statistics.

Warren's test may be couched in science, but its ideological force stems from his conservative Christianity. As the spokesman for eHarmony, Warren appears in many commercials and the site prominently displays his image, academic credentials, and clinical psychology experience. The site lists Warren's "expert guidance that underlies everything we do" as "one of the greatest benefits of using eHarmony."[88] He himself has underlined his and eHarmony's paternalistic attitude by stating, "We do try to give people what they need, rather than just what they want."[89] Warren's views on relationships are rooted in Christian values including abstinence, as he argues both that couples should wait two years before getting married and that premarital sex "clouds decisions."[90] Warren has also disparaged premarital sexuality throughout his writing and thinks of eHarmony as a tool to combat the lack of Christian values that "the so-called 'cohabitation epidemic'" connotes.[91] Warren designed eHarmony's matchmaking program to restrict dating choices based on his particular moral conception of what a relationship should consist of and advertises this program by implying that his way of creating relationships has been scientifically proven to be the best.

A lengthy YouTube commercial shows a day in Warren's life as he goes to work at eHarmony, attends meetings, and talks to people in offices and cubicles.[92] He discusses his wife and narrativizes his professional history and desire to use eHarmony and his personality test to lower the divorce rate and the number of unhappy marriages more generally. This and similar ads present Warren as employing a hands-on approach at eHarmony, encouraging users to think of him as the actual matchmaker behind the algorithms. In turn, these ads declare the importance and centrality of the personality test and resulting profile to eHarmony's business and advertising strategy. To back up Warren's claims, eHarmony hired Harris Interactive to survey its user base and continually publishes its results, along with findings from a peer-reviewed article in the 2012 *Proceedings of the National Academy of Sciences* and other sources that show not just that eHarmony produces more marriages than any other site but also that these marriages tend to be more satisfying and lead to fewer divorces than those produced via any other meeting place, online or off.[93]

While these statistics are laudable, the inextricable connection between Warren and eHarmony as a company became a hindrance in 2007 when many journalists began to criticize Warren's religious views and ask whether eHarmony had a Christian agenda. Before starting eHarmony, Warren was not only a marriage counselor but also a seminary professor and Christian theologian. He had also closely associated himself with James Dobson, the founder of the evangelical activist and media production group Focus on the Family (FOTF), which promoted conservative Christian values and opposed LGBT rights, along with a host of other "vices" that its leadership felt undermined traditional Christian family values. Warren regularly appeared on Dobson's radio show and published many of his self-help books through FOTF. Although eHarmony had always been open to a secular audience, it was originally marketed as "based on the Christian principles of Focus on the Family author Dr. Neil Clark Warren."[94] Warren had also conceived of his company in evangelical and Christian colonial terms and in 2003 took his business global in the hopes of serving missionaries.[95] While, early on, 85 percent of eHarmony's members were Christian, Warren never wanted the site to be exclusive and instead argued that "sometimes non-Christians need Christian efforts" like his.[96] While this marketing helped eHarmony gain an audience early on, by 2007 it had become a drawback as the company tried to gain more members by reaching out to more secular audiences.

By 2005, ostensibly due to changing attitudes toward LGBTQ rights, Warren had stopped appearing on Dobson's show, bought back the rights of his books so that they would not have FOTF on their covers, and erased any mention of specifically Christian values from eHarmony's website and advertising.[97] However, eHarmony refused to allow homosexuals to use its site for matchmaking, claiming that it only tested its algorithms on heterosexuals and therefore could only match these kinds of relationships. Many on both the right and the left read this as "transparently convenient."[98] From the right, Warren received questions like, "The site says that eHarmony doesn't offer homosexual matching because you have 'no expertise when it comes to same sex matching.' How do you defend that politically correct answer?" He responded by emphatically stating his opposition to homosexuality but explained, "Cities like San Francisco, Chicago or New York—they could shut us down so fast. We don't want to make enemies out of them. But at the same time, I take a real strong stand against same-sex marriage, anywhere that I can comment on it."[99] Warren presents his scientific rhetoric about the differences between heterosexual and homosexual relationships as a conciliatory cover for his bigotry.

Moreover, this answer indicates eHarmony's larger efforts at using science in order to prop up Christian values. Academics and journalists alike have long been skeptical of eHarmony's scientific claims since its representatives rarely present data at conferences and never publish their proprietary methodologies or findings.[100] Social psychologist Benjamin Karney argued that eHarmony was duping customers and said its "scientific methods" were "basically adorable."[101] In turn, Warren uses statistics on the happy marriages that eHarmony has produced as proof of eHarmony's scientific validity. Rather than focus on the experiments that led to the creation of eHarmony's matchmaking software, Warren uses these statistics to frame eHarmony itself as what journalist and self-help writer Lori Gottlieb calls "the early days of a social experiment of unprecedented proportions, involving millions of couples and possibly extending over the course of generations."[102] In the process, Warren's conservative Christian traditionalism, which he now refers to as a secular "folksy wisdom," becomes the hypothesis that eHarmony works to prove.[103]

Those on the left rejected Warren's "scientific" claims as the basis for excluding homosexuals, and several parties sued. As a result of settled discrimination lawsuits in New Jersey and California, in 2009 eHarmony created Compatible Partners, a same-sex dating site. Like eHarmony, Compatible Partners specializes in creating "lasting relationships," and the site includes a personality test and matchmaking program that are virtually identical to eHarmony's and that privilege relationships built on gender binaries.[104] As it has always done with Jewish, black, Hispanic, and Asian people (as well as seniors), eHarmony now redirects users interested in same-sex dating to its partner site, though it does very little to advertise this service and mentions it only at the very bottom of its homepage in small type. Ironically, Cupid's Library, a prominent review site for online dating services, criticized Compatible Partners for being too similar to eHarmony and for assuming that a dating service for heterosexuals would work for homosexuals as well.[105] Even so, two hundred thousand people signed up for the site within the first year.[106]

eHarmony makes dating purposefully complex, laborious, expensive, and time consuming in order to both attract users highly motivated to find a long-lasting relationship and make the resulting dates as promising as possible. As Marylyn Warren, the senior vice president of eHarmony and wife of Warren, argues, "If you want a date for Saturday night, we probably are just not the place to come."[107] As previously mentioned, beyond the lengthy test itself, eHarmony up until recently also regulated how users could first contact each other and offered specific topics for them to discuss.

To create its monogamous, heterosexual matches, eHarmony originally administered an exhaustive, hours-long, multiple-choice personality test that included 436 questions. This test, which eHarmony reported fifteen thousand people took every day in 2013, asked users about everything from the cleanliness of their rooms to their interest in beaches.[108] Reportedly based on Warren's clinical experience, eHarmony's test matches users based on its trademarked "29 Dimensions of Compatibility," which heavily stresses the need for similarity between the personalities and activities of partners. Buckwalter has stated that "it's fairly common that differences can initially be appealing, but they're not so cute after two years. If you have someone who's Type A and real hard charging, put them with someone else like that. It's just much easier for people to relate if they don't have to negotiate all these differences."[109] While studies have used the Big Five traits to examine the relationship between different personalities and job or school performance, eHarmony instead uses and studies (via its eHarmony Lab) the Big Five traits as they apply to the creation of relationships.[110]

While Chemistry.com's focus on compatibility can lead to a gendered stereotyping, eHarmony's focus on similarity can lead to various forms of segregation. This segregation is not necessarily focused on ethnicity or culture, but it does ensure that diversity is seen not as a value but rather as detrimental to relationships. eHarmony's recommendation system is based on studies that show that people's preferences "were most strongly same-seeking for attributes related to the life course, like marital history and whether one wants children, but they also demonstrated significant homophily in self-reported physical build, physical attractiveness and smoking habits."[111] eHarmony's recommendation system primarily looks for these similarities, and its goal is therefore to "find somebody whose intelligence is a lot like yours, whose ambition is a lot like yours, whose energy is a lot like yours, whose spirituality is a lot like yours, whose curiosity is a lot like yours."[112] As Warren put it, "These are the two principles we believe in: emotional health and finding somebody who's a lot like you."[113] While these practices are not necessarily racist or classist, this focus on similarity in areas of interests, hobbies, and lifestyles that are highly racialized and class based can lead to an automated segregation in the relationships eHarmony creates.

Many of the rest of eHarmony's questions are quite ambiguously phrased, which makes it difficult to tell exactly what effect the answers will have on generating matches. For instance, answering "very important" to the question, "How important is your match's age to you?" could result in the test's eliminating all matches that are older than the user, or all matches that are not the same age as the user, or matches that are over a particular age regardless of the age of the user. eHarmony asks similar questions about the importance of a match's education, religion, income, and ethnicity. Several of these questions make stereotypical assumptions about users based on class, gender, and other social factors. In turn, they force users to think of themselves through this stereotyped vision. Users cannot state that they are interested in someone who is labeled as unattractive, uneducated, or has a low income; they instead must state that they do not care about the level of their attractiveness, education, or income at all and hope that some of their matches fit the criteria they desire. The ambiguity and limitations of these questions both limit the ability of people to describe themselves adequately or accurately and present them and their desires in a simplistic fashion.

Users are then given a five-page list of adjectives and are asked to state how well the words describe them, from "not at all" to "very well." The list appears random but includes many opposites like "warm" and "aloof" or "predictable" and "spontaneous." At the end of this list of adjectives, users must state which four adjectives their friends would use to describe them. This section both encourages users to think of themselves from an "objective" outside perspective and tests users on how well their sense of self maps onto how they think others see them. Much of the rest of the test presents users with statements, social values, activities, and skills. Users rate how well they identify with the statements and social values and state their level of interest in and knowledge about the stated activities and skills.

Through these questions, the computer generates a profile used to match the user with others. These questions also break users down into what Illouz calls "discrete categories of tastes, opinions, personality and temperament."[114] This compartmentalization is read back to users and proposes both how they should self-manage and who they might successfully date. This process of filling in bubbles turns intimate life and emotions into "measurable and calculable objects, to be captured in quantitative statements. To know that I score a ten in the statement 'I become anxious when you seem interested by other women' will presumably lead to a different self-understanding and corrective strategy than if I had scored two."[115] The use of these sliding scales only adds to the difficulty of both knowing how to answer them "correctly" and figuring out what the answers say about the user in question. While the difference between a zero and a ten is clear, it is much less obvious what a two compared to a three or four might signify to the personality test algorithms.

While it is common for matchmaking sites to state that their profiles are generating self-reflection by helping "you learn more about yourself and your ideal partner," they also shape users and teach them how to manage their online representation.[116] With its attempt to minimize sexuality in dating while it continually urges users to describe themselves truthfully, eHarmony demonstrates what Michel Foucault called "the link between the obligation to tell the truth and the prohibitions against sexuality," which he argued was "a constant feature of our culture."[117] Concomitant with this logic, eHarmony's test implies a direct link "between asceticism and truth"; eHarmony argues that only by taking sexuality out of the equation (at least when finding a date) can users "decipher" themselves as sexual beings.[118] Its test acts as a "technology of the self" designed not only to "transform" users "in order to attain a certain state of happiness, purity, wisdom, perfection, or immortality" but also to "determine the conduct of individuals and submit them to certain ends or domination."[119] It collapses any distinction one might make between what Plato called "concern for oneself and the knowledge of oneself"; between what we might now call "self-reflection" and "self-management."[120] While self-reflection ostensibly is geared toward gaining a deeper awareness of oneself and how one appears to others, self-management highlights the desire to transform and restrain oneself in order to be more acceptable and successful. How can users learn about themselves if they do not even understand the questions being asked or have an answer that is more complex than "yes" or "no"? Instead, these tests and resulting profiles are designed to shape users with the goal of making them desire (and be more desirable to) the other users on the site.[121]

The resulting personality test profile is often what Paumgarten calls a "vehicle for projecting a curated and stylized version of oneself into the world" that signals "one's worth and taste."[122] Some compare the process of writing a profile well to "showing up in a black Mercedes."[123] This comparison between writing a profile and displaying your consumercentric riches is an important and telling one, as it displays how this self-management, like so much of postfeminist culture, is deeply connected to the same rhetoric and discursive strategies of consumerism more generally. Indeed, dating sites rely on consumerist rhetoric and skills throughout their recommendation process.

This Machiavellian process of self-management forces users to express their "capacity to be split between the private and public realms of action, to distinguish and separate morality and self-interest and to shift back and forth from one to the other."[124] On these dating sites, the diametrically opposed rhetorics of self-reflection and self-management overlap and merge with each other. While self-reflection often presupposes a Cartesian sense of a unified, essential self, self-management articulates a very different type of postmodern self that continually shifts as the economic situation and culture around it shifts. Dating sites advertise themselves as a tool of self-presentation that "presupposes a movement inward toward one's most solid sense of self" rather than outward toward possible others.[125] Personality tests and profiles are not created for what Illouz calls "a concrete, specific other, but for a generalized and abstract audience," allowing users to explore their core self and discounting the idea that the self is actually simply a "multiplicity of roles to play."[126]

The result of these self-management techniques is that while online dating sites stress that their tests are designed to find what makes a user a distinct individual, they heavily value conformity because it makes it easier for their algorithms to find potential matches. While scholars of postfeminism and neoliberalism focus on how self-management is now practiced at the level of the individual in ways that diminish the communal ties that feminism largely worked to build, this is complicated by online dating's insistent push for conformity. These matchmaking algorithms and filters make it difficult for anyone out of the "ordinary" (often defined on these sites as their preferred customer base of white, middle-class professionals) to be matched on many sites, as these sites tend to introduce similar people to each other. Outliers, especially in relation to income and education, are more difficult to find matches for, and eHarmony often simply denies them access to its site.

Yet, for many users, no personality profile is provided, and instead they are turned away from the site entirely. Along with questions that determine a user's sense of self, the test includes a validity scale designed to assess whether users are lying about themselves or answering questions randomly. This section also includes various more pointed adjectives, including "depressed," "hopeless," and "out of control," which are meant to determine whether users have dysthymia, or chronic but not necessarily severe depression. These scales should be viewed as only the latest attempt in a long history of trying to make computerized dating appear to be a safer alternative to meeting someone in a bar. In addition, eHarmony also advertises an identity theft protection plan, as well as other services that help guarantee that the person that you are talking to online is not trying to deceive you.

If your scores on eHarmony's validity and dysthymia scales indicate you may be lying, have depression, or are otherwise undesirable, eHarmony will reject you from its service. These users can still see their personality profile, but they are first taken to a page stating, "We're very sorry, but our matching system cannot predict good matches for you."[127] In 2007, 20 percent of those who took the eHarmony test were rejected due to their answers. The site does not give any specific reasons to the users regarding why they were specifically rejected, but statistics about these rejections are available. One-third of those rejected are currently married or separated.[128] One-fourth of those rejected are younger than the site's minimum age of twenty-one, an age cutoff to which eHarmony adheres because 85 percent of marriages before the age of twenty end in divorce.[129] I was rejected by eHarmony and had no idea why until I read that it also rejects people with PhDs because there are not enough people on its site with commensurate levels of education to match us with (at least that is why I would like to believe I was rejected). eHarmony also rejects people under sixty who have been married more than four times. Another 9 percent are blocked for giving inconsistent answers on the test that imply they were not paying attention, were lying, or have psychological issues. The company states that it only screens out severe depression, but dysthymia scales are not very accurate and screen for any kind of depression, which Warren equates with both "low energy" and "a lack of emotional health."[130] It also screens for obstreperousness, as it links this characteristic to being hard to please and unable to enjoy long-lasting relationships. Warren explained these rejections by stating, "You'd like to have as healthy people as you can. We get some people who are pretty unhealthy. And if you could filter them out, it would be great. We try hard. And it's very costly."[131] With this eugenic logic, recommendations filter users not only from the site but from the gene pool entirely. eHarmony does not tell such users that they have been deemed too mentally ill to be part of the dating pool and rejects them while also suggesting they study their personality profile for "valuable insights" to help in improving themselves.[132] Rather than encourage these users to see a psychologist, eHarmony directs them toward its secondary business of providing self-help advice designed to give them lessons on how to help themselves.[133] In 2007 *Time* emphatically stated that eHarmony's actions constituted discrimination and labeled it one of the five worst dating sites, stating, "Our main beef with this online dating site is its power to cause utter despair."[134] In its bid to produce successful, monogamous relationships, eHarmony in fact passes judgment about who appears to "deserve" a relationship and who does not.

Parody and Algorithms of the Oppressed

Online dating sites—and eHarmony in particular—argue that customers are incapable of finding a good relationship on their own and that some are not worthy of having a relationship at all. Yet there has also been a consistent critical response to dating sites that largely works to reaffirm the agency and ability of people to make up their own minds about who they should or should not date. In these often quite creative responses, which range from parody to satire to thoughtful critiques, we can see a challenge not just to the idea that algorithms are smarter than us but also to the idea that algorithms are a domineering, if not entirely unstoppable, force that can transform our culture whether we want it to or not.

Parody has become an especially prevalent form of response to online dating. As Linda Hutcheon has argued, parody "allows an artist to speak TO a discourse from WITHIN it, but without being totally recuperated by it." As a result, it has become "the mode of the marginalized, or of those who are fighting marginalization by a dominant ideology."[135] Yet many of these responses have also come from within the online dating industry itself. For instance, one of Chemistry.com's first ad campaigns features photos of various puzzled-looking people with the words "Rejected by eHarmony" stamped over their faces in red. The ad then states that you should not feel bad, since millions of other people have also been rejected and there is no way of knowing why. Chemistry.com instead tries to turn this rejection into a badge of honor by asserting that "those diverse, keep-you-on-your-toes people" are the customers being rejected, whom Chemistry.com would gladly accept. One notable version of this ad features a young black man in a hoodie asking, "Was it my hobbies?" The ad does not answer this question, but by pointing to the diversity of users that eHarmony rejects, it does imply the fear that eHarmony's algorithmic rejection process is merely a smokescreen for racism and other harmful practices. Another of these ads features a gay man considering his rejection, which more directly points to how eHarmony uses its algorithmic rejections to obscure its homophobia.

Though certainly critical, Chemistry.com's response is also extremely self-serving. While attacking eHarmony's algorithms, it presents its own as a model of utopian diversity and acceptance. This rhetoric still feeds into the fallacy that there is one algorithm out there that can guarantee that you will find true love—and that it is the one. Yet there are many critiques that take a more aggressive stand against online dating more generally and the very idea that any algorithm could accurately match people together or that any database could store and quantify what makes us unique.

Throughout the mid-2000s, along with far too many *Saturday Night Live* skits, several parody dating sites emerged that illustrated both how normalized online dating had become and the many anxieties that still surrounded them. They often made fun of how online dating sites present their algorithms as omnipotent and beyond scrutiny. While real sites portrayed these algorithms as what made them unique, parody sites portrayed them as merely a ruse designed to lure potential customers. For instance, SuperHarmony.com, a site for those interested in dating superheroes and villains, features a nonfunctioning short "superQuestions" compatibility test that asks about your interest in superheroes versus supervillains and various personality test questions, such as whether your home planet exploded, what your favorite superpower is, what radioactive animals have recently bitten you, and whether you are "the last hope for your race/species/planet." The test mainly includes simple yes-or-no questions and five-point response scales (from strongly disagree to strongly agree) that echo eHarmony and Chemistry.com in both appearance and content. The disjuncture between these questions and the format of the test pokes fun at the idea that any multiple-choice test could accurately define the type of person you are or help you find love.

Other parody sites highlight how matchmaking algorithms can facilitate racist and otherwise vile couplings. A fake commercial for the nonexistent OKMudblood site presents the possibility that such tests might match you up with Voldemort or Bellatrix Lestrange. In the process, it harks back to long-running anxieties that computerized and online dating may expose us to dangerous and nefarious people and situations. In a similar vein, a commercial for the fictitious site Whites Only Dating states it uses its compatibility test to match "light pigmented" people with others with a compatible white skin tone. To entice new users, the commercial advertises that "as a part of your free membership, you can take a journey of self-discovery with the color-coded personality test."[136] This eHarmony-esque test uses information about your likes and dislikes to generate a profile declaring which shade of white you are, from pure white to a yellowish hue the site refers to as papaya whip. In a parody of most personality tests, the commercial shows a profile page declaring, "You are Papaya Whip!" which bestows the person with "the self-reliance and assurance of a typical white person." This test links ideals of racial purity to an algorithmic desire to continually capture and compute an ever more precise image of the world and ourselves. While technology companies typically frame this desire in utopian terms as a way to make the world more sustainable and orderly, this commercial also makes links to both the eugenic projects of Galton and the meticulous record keeping of the Nazis.

The page goes on to say, among other things, that this pigment gives people a sense of purpose and makes them "equitable, intellectual and shrewd and may be [sic] a bit of a workaholic." Such characteristics could be pulled from any of a number of real, automatically generated dating site profiles (or horoscopes, for that matter). Such profiles are vague enough that they can be applied to most anyone but include enough specific details that anyone can imagine they are speaking directly to them. The site's focus on whiteness satirizes the conspicuous racism that pervades much of online dating culture. Connecting a personality to a pigment is, on its face, ridiculous but is also central to how stereotyping works. The commercial features classical music, which is heavily associated with affluent whiteness and illustrates how our understanding of cultural interests shapes our perception of race. This test may be a parody, but it does help us examine how race and racism function in digital culture and through auto-

mated algorithms. Personality and taste information can quickly and inap-
propriately be used by recommendation systems as a proxy for race not just on
dating sites but across digital industries. Indeed, just as we construct our understanding of race and ethnicity through a complex—if often also harmful and
misleading—understanding of taste, personality, class, and a myriad of other
aspects, this parody implies that online dating sites take a much more simplistic and dim-witted tack that both takes in much less information and often generates much more static and harmful portraits of users.

While these parodies illustrate the various problems of online dating and
matchmaking algorithms, they also unfortunately make clear how limited these
efforts at transforming digital or algorithmic culture are. For instance, rather
than tamp down the racism of online dating, Whites Only Dating seems to
instead often inspire these practices. The YouTube comments under the video
primarily express sadness that the site is not real, consolation that "whitepeo-
plemeet is practically every dating site," and anger that there can be dating
sites specifically for African Americans and Latinos but not for Caucasians.[137]
Sadly, similar, real sites have begun sprouting up, such as WhereWhitePeople
Meet.com.

When Razmig, the creator of the Whites Only Dating video, explained
why he had made the site, one person commented, based on his picture, that
he thought it was ironic that he made the site since he was not white but rather
Arab.[138] Razmig then countered by stating that he actually was white, and
they then engaged in a protracted argument over exactly how white he was,
with the commenter calling him a "mutt" at one point. This argument ended with
Razmig's proclaiming that his family was from Central Eurasia, which appeared
to settle the argument but also illustrated how contested race can be online.
Razmig actually appears to be adopting the same logic of racial specificity that
Whites Only Dating parodies. In a space where race is rarely known for sure,
it clearly does not disappear but rather is constantly negotiated and constructed.
One other responder to Razmig argued that a "whites only" dating site would
not be offensive since every other group has one and "dating is too personal"
to be thought of as a domain for racism. Furthermore, she asserted, "I don't
want to spend a lot of time sifting out individuals who belong to groups I am
not interested in and I doubt most of those individuals want to date me
either."[139] This response illustrates how algorithmic culture connects racism to
a capitalist logic by framing it as simply a rational, time-saving, efficient, and
productive exploit.

While these efforts do not appear capable of transforming online dating sites
themselves, there have been efforts to at least make it possible for people to take
advantage of them while avoiding some of their more problematic aspects. One
particularly popular browser extension made by Ben Jaffe, "OkCupid (for
the Non-Mainstream User)," changes the interface of OkCupid in order to

highlight how potential matches answered questions concerning (among other things) whether they discriminate based on race or other physical traits, what kind of relationship they are looking for, and what their dietary restrictions are. While this extension helps users screen for those who answered their questions in racist, ableist, and transphobic ways, Jaffe specifically created it to "expose the question data in a more clear and filtered way than OkCupid does." He explains, "OkCupid boils all of that nuanced data down to a match percentage on the profile. If you clicked into the questions area, you could see more, but it was still very high-level. I built the plugin for myself to find people who are science-minded, kind, and compatible with my particular relationship paradigm (consensual non-monogamy)."[140] User reviews for this extension also point out that while it is quite buggy in the way that many amateur-created programs are, it helpfully transforms OkCupid from a site that primarily highlights stereotypically mainstream, heterosexual users into one that emphasizes the diversity, surprise, and serendipity that matchmaking algorithms tried to eliminate from their computation.

Yet working around OkCupid's interface is not a great solution either. Jaffe, as a white, cis-gendered male, felt uncomfortable "recording, in code, how good/bad each answer was for classes of questions around things [he had] no direct experience with." He explains, "I think of myself as an empathetic person, but I certainly don't want to be the arbiter of what counts as discriminatory. In some cases (sometimes in response to me reaching out), members of those communities have chimed in."[141] No matter your intent, trying to work around an algorithmic system with more algorithms causes problems, as their programming still necessitates a degree of binary thinking about what exactly a good answer and a good relationship consist of. Data scientist and activist Cathy O'Neil argues that people trust in algorithms and think of them as objective and scientific as a way to "avoid difficult decision-making, deferring to 'mathematical' results."[142] She uses the example of the personality tests that many supermarkets and other major stores use to screen job applicants. Much like eHarmony, these tests rely on Big Five personality trait questions to figure out who would make a good employee. If you have a personality that the test creators decided would not make you a good employee, you simply never get called for an interview and are not told why. In such cases, it is clear that these stores use algorithms to avoid the hard work of really considering what makes a good employee.

Yet Jaffe's decision to create his own extension and his consideration of the ethical quandaries he faced while programming it illustrate that these algorithms can also drive you to confront difficult decisions head on, especially when you become aware of how unfair they are. Whether in the case of matchmaking systems, employment personality tests, or the algorithms that tell airline attendants which passengers to kick off first, these deferrals to the ostensible objectivity of algorithms can make us keenly aware of systemic injustices and perhaps drive us to a more critical outlook and the realization that the questions that we push onto algorithms are often the ones that have no right answer at all.

While it is great that online dating can bring people together who may never have otherwise met each other at all, too often these sites try simply to re-create the most problematic parts of normal dating and the relationships to which they often lead. The creators of these sites argue that their matchmaking systems can lead to better marriages and families. Yet their idea of what a better family consists of is influenced by eugenic technologies and rhetoric that only supports homogeneous couplings; on these sites, difference is only valued insofar as it supports the heteronormative gendering of relationships.

That is not to say that people cannot challenge these systems and use them to their own ends. Presently, their attempts and approaches are largely just bandages over extensive social injustices. Yet the examples I have discussed also point to the variety of ways people interact with online dating sites and the heavily negotiated ways they think of their relationship to their matchmaking algorithms. They reveal that these algorithms may often be viewed less as the fundamental aspect of online dating and more as a tool that many instead must continually work around and in opposition to in order to actually find, meet, and get to know their potential loves.

5

The Mirror Phased

• •

Embodying the
Recommendation via Virtual
Cosmetic Surgeries and
Beautification Engines

Users often view algorithmic technologies as mirrors that merely reflect them and their desires back to themselves. Yet, as Jacques Lacan famously argued, when we look in a mirror, we do not see ourselves as we are; we may identify with our reflection, but we, as fragmentary, mutable, and multifarious subjects, can never quite measure up to the image of wholeness, cohesion, and mastery that it presents. Here, I consider how we come to view ourselves through algorithmic technologies by examining how cosmetic surgeons and cosmetics companies use them to recommend surgeries and makeovers to patients and users. Even though I feel the world does not actually need more "gazes," I am tempted to call the way users view themselves through these technologies as an algorithmic gaze. This gaze does to bodies roughly what Netflix and Amazon do to taste and teach us to see ourselves as if algorithmically generated and modular by design.

For example, on Anaface, users upload images of their faces and then click on specific points of reference, such as the bottom and top of their earlobes; the edges of their noses, mouths, and eyes; and so on. The program then uses these points to, as the site eloquently states, "score your face" on a scale from

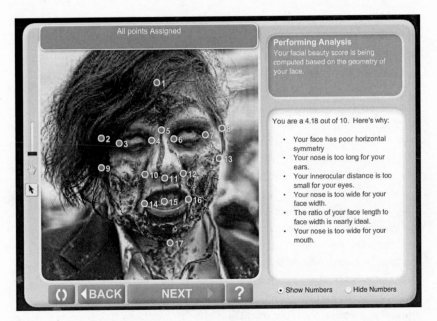

FIG. 5.1 Screenshot of Anaface

one to ten.[1] The press release for the site's launch explains that the tool is designed to help users "know what features of [their] face detract from [their] facial beauty [so that they] can make corrections" ranging from wearing a different type of glasses to deciding on a particular cosmetic surgery.[2] It also argues that advertisers could use the site to select models and surgeons could use it to "show potential patients how their looks could be enhanced by changing the shape of their eyes, nose or ears."[3] Within these contexts, Anaface's primary purpose is to generate recommendations for making users as beautiful as possible.

Anaface does not simply provide users' numerical scores; it also lists everything that is positive and negative about users' faces. The program focuses on issues of symmetry and the ratios of different facial features to one another. While it may say, "Your face is too narrow/too long," or, "Your mouth is too wide for your nose," it may also try to soften the blow by stating, "The ratio of your mouth width to nose width is nearly ideal."[4] For those considering surgery, these comments can provide the basis for deciding what types of surgeries to get.

Digital imaging systems fragment the body into isolated and generic parts so that it becomes a vehicle for continual upgrades. As these technologies train subjects to view their bodies as endlessly pathological, these bodies come to (seemingly) require endless modifications. Bernadette Wegenstein argues that the cosmetic surgery industry more generally teaches subjects to view their own and others' bodies as "incomplete projects" and encourages these subjects to use technology and cosmetic surgeries to reveal their "real" or true self.[5] Similarly, Alexandra Howson, like Kathy Davis before her, shows that patients view cosmetic surgery and its concomitant technologies as tools to individuate themselves in ways that treat "the body as a vehicle for self-expression."[6] They do not view the body as naturally expressive but instead believe it can become so through modifications; rather than an essential aspect of the self, the body here becomes a Foucauldian technology of the self. Paradoxically, as Anne Marie Balsamo has noted, while patients decide which operations they would like, these modifications, whether a new nose, breasts, or stomach, do not individuate but rather "reconstruct the body to cultural and eminently ideological standards of physical appearance."[7]

Thus, Anthony Giddens has stressed how the body is now "a phenomenon of choices and options."[8] This representation of identity becomes linked to "the consumer choices one makes to 'enhance' it."[9] It is not only cosmetic surgery that makes bodies into a technology of the self but the entire apparatus of recommendations that surround them as well. In the case of the cosmetic surgery industry, a focus on recommending the "right" personal choices—those designed to achieve the most empowerment via economic and social gains—has led to bodies that display conformity rather than distinctiveness. Kathryn Pauly Morgan has referred to this phenomenon as a "paradox of choice" wherein "what looks like an optimal situation of reflection, deliberation and self-creating choice often signals conformity at a deeper level."[10]

As problematic, if not outright objectionable, as it may be to undergo a major surgery solely to conform to ideals of beauty better, these programs do not necessarily have to be used to generate recommendations. Outside the clinical context, they can simply be used for fun or even to make light of patriarchal ideals of beauty. Placing coordinates incorrectly or attempting to calculate the beauty of your jack-o'-lantern can be far more gratifying than trying to follow the sites' instructions. Briefly, a meme erupted out of the Anaface images as people began rating the faces of everyone from Grumpy Cat to Jesus. Collectively, these images, which give the Guy Fawkes mask roughly the same score as Ryan Gosling, dispel any notions that these ratings are either scientific or accurate. The hundreds of images associated with this meme also illustrate the various resistant responses people have had to Anaface and its underlying ideology. While many uploaded images of seemingly attractive celebrities like Brad Pitt and Jennifer Lawrence only to have them deemed average by Anaface, many others used the technology to evaluate Adolf Hitler and animated characters such as Shrek, whom Anaface assigned oddly high scores. These uses of the site challenge Anaface's assumptions that there is an essential way to judge beauty and that one's looks are related to one's self-worth.

Even so, simply seeing your physical body as potentially endlessly changeable can be empowering and pleasurable. The resulting image, plotted out with coordinates, depicts your face as plastic and designed to be transformed and manipulated. While Lacan argued that identifying with images of ourselves can lead to a sense of cohesive wholeness and mastery, Anaface and programs like it instead encourage us to imagine ourselves as fragmentary, incomplete, and multiple. In the context of cosmetic surgery, seeing ourselves within a grid covered by intersecting coordinates does not make us appear static and contained but rather highlights how mutable we are. These coordinates, after all, are meant to be moved.

Through the use of these coordinate systems, programs like Anaface encourage us to see ourselves from the perspective of a digital recommendation system. Just as all of the recommendation systems I have discussed transform our personalities into fragmentary data points that are constantly rejiggered and compared to others in order to generate a continually transforming encapsulation of who we are and who we want to be, these digital imaging programs attempt to do the same thing with our physical bodies. While several of these programs use similar recommendation technologies to those discussed in previous chapters, others work more broadly to recommend new bodies through peer pressure and in the context of patient-surgeon relationships. Even so, discussing these imaging technologies in relation with these other culture industries of choice elucidates how they together teach us to see ourselves (and specifically our bodies) not as inherent and essential objects but rather as a technology of the self; paradoxically, a product of both disciplinary control and individual choice.

To add context to these technologies and how cosmetic surgeons currently use them in their practices, I begin by explaining how photography and "before-and-after" shots have historically been used in clinical settings. Surgeons have long used photography to help give patients a sense of how they look from different, unfamiliar angles so as to simulate an "objective," outsider perspective. The amassing of standardized before-and-after photos helped to structure cosmetic surgery as a medical field with specific measurements of aesthetic success and failure. In the process, particular ways of viewing patients' bodies also became normalized and these patients were taught to judge themselves from unfamiliar perspectives, as "others" to themselves. These efforts taught patients to take on the surgeon's perspective—and that of the dominant culture—when judging their own bodies.

I then compare this history of cosmetic surgery photography to the current use of digital imaging technologies. Many surgeons now feature such technologies on their websites as lures to entice potential patients to their clinics. In many cases, companies present these technologies as a chance for users to playfully explore and reshape their faces and bodies, while also asserting that digital modifications can be achieved through actual surgery. These technologies thus domesticate cosmetic surgery by presenting it purely as fun and spontaneous. This movement from the clinic to the home in some ways gives people more control over how they can imagine their new bodies but can also simply further the normalization of seeing themselves only as a site for constant critique and correction.

I then discuss how these technologies have begun to be used to algorithmically recommend particular bodily changes and sometimes automatically change users' images to make them appear more beautiful in the way the system has determined is best. These programs, like Anaface, base their ideas of beauty on a wide range of classical attempts to define it, as well as modern collaborative filtering and artificial intelligence methods. These technologies treat attractiveness as a scientifically measurable, objective phenomenon that transcends history and culture. I then consider how self-help companies have taken advantage of this aspect by using these technologies to recommend lifestyle changes. These companies create programs that show users what they will look like later in life if they continue to smoke, eat unhealthily, or suntan. Users input information about their lifestyles, and these programs automatically modify their images to show how their present life choices will affect the way they look ten, twenty, or thirty years later.

Throughout this chapter I examine how digital imaging technologies mediate and structure relationships between people and their bodies in both medical and domestic spaces. Cosmetic surgery is often framed either as an oppressive form of control or as a liberating expression of empowerment. This debate largely centers on the role of choice and recommendation in patients' decisions about both whether to get surgery and what changes they would like to make: Are these choices an empowering expression of the patient's individuality, or are they based on heavy-handed recommendations from popular culture and their surgeons that patients willingly accept? By focusing on the digital imaging technologies that mediate the patient-surgeon relationship, I add nuance to this debate by showing how this apparatus paradoxically constructs cosmetic procedures as an individuating and empowering experience for the patient even as the surgeon remains the center of power and authority.

These programs construct the choices and expectations of users in ways that simultaneously encourage them to experiment with their virtual bodies as an expression of personal fantasy and unlimited choice while also continually recommending certain dominant, culturally determined options over others. As in other areas of the culture industries of choice, users can experiment and choose from a variety of options, but these technologies always privilege certain recommendations over others. Ultimately, these recommendations affect both how people feel about and imagine their bodies and—if actual surgeries are involved—how they will look in the future.

Photography and the Standardization of Plastic Surgery

While companies have only recently begun using digital recommendation technologies to judge the physical body, the practices that surround this use arise out of a much longer history concerning the intersections between visual media and medicine, intersections that have helped to define beauty as a sign of health. Much of this stance rests on the logic that our sense of attraction is governed by our search for healthy and fertile mates, but this neglects to take into account how varied attraction can be and overlooks the fact that fertility has little to do with body shape.[11]

The use of digital imaging technologies by cosmetic surgeons has developed out of a long history of surgeons' using photography to standardize their conception of beauty (and thereby what a successful surgery consists of)—a necessary step in turning their field into a venerable profession. In the mid-1980s, photography, due to its ability to document and record the preoperative and postoperative body, helped create standard practices and objective criteria for the profession. These photos could be used to create specific expectations for future patients and also teach medical students what the proper outcomes of these surgeries were. Moreover, basing the success or failure of a cosmetic surgery operation solely on the surgeon's and patient's memories of the patient's former body was deemed unacceptable, as the surgeon and patient might remember this body differently or have had different expectations of what was possible through a surgical operation. At the same time, certain "defects," such as asymmetries in facial features, that were invisible before surgery might become more noticeable after a surgery, and a photographic record could provide proof that the defects, while perhaps unnoticed, had always been present. Before-and-after photos provided evidence of whether these surgeries were ultimately successful, which is important for both professional and legal purposes. As cases in which the patient is unhappy with the surgery outcomes can often end up in court, these photos could provide objective proof of whether the surgery should be considered a success.

At the same time, these photos encourage patients to view themselves from the perspective of the technologies used to create the images and imply that what they look like in visual media is even more important than how they look in person. This logic has only become more prevalent today as we spend less and less time communicating in person. Since a subject may look wonderful from a high-angle medium shot through a 35 mm lens but terrible in a low-angle close-up with a 15 mm lens, photography had to become standardized for this medical field.[12] Cosmetic surgeon and University of California, Los Angeles, School of Medicine instructor Robert Kotler has argued that since different lighting, angles, and expressions can affect how a person looks as much as any surgery, before-and-after photos must be "technically consistent" so patient-consumers "can make an honest judgment" of a surgeon's artistry.[13] Instead of hiring professional photographers to document patients and surgeries, doctors or their assistants normally took these photos and therefore required some training in the techniques and technologies of this artistic field. In 1988 Gerald D. Nelson and John L. Krause, for the Clinical Photography Committee of the American Society of Plastic and Reconstructive Surgeons, prepared *Clinical Photography in Plastic Surgeons*, a landmark collection of essays that served to codify the best practices in clinical photography. In this textbook, surgeons and lawyers both historicized the long history of medical imaging (stretching back to Albrecht Dürer but focusing on the beginning of photography and the interest in "documenting deformity, disease, and injury") and gave detailed descriptions for doctors on how to photograph surgical subjects.

To provide the greatest evidentiary value and create a sense of objectivity around cosmetic surgeries, every element of photography had to be standardized in order to allow this largely qualitative evidence to be generalized and turned into quantitative statistics, the lingua franca of contemporary medicine. The authors in *Clinical Photography in Plastic Surgery* worked to create a set of best practices concerning how to standardize their photography as much as possible. This textbook first went over the types of cameras, lenses, lighting, and film stock to use when photographing a nude body at a distance, close up (with special focus on shots of the inside of the mouth), and in an operating room. Later articles then went on to address how best to project these images, how to store them, and the special ethical considerations concerning these photos in relation to patients' privacy concerns and the fiduciary duty of physicians. In the process, these articles provided a framework for the standardization of the relatively new profession of medical photography and the types of bodies that can be recommended.[14]

As each element of the photographic apparatus will distort the body in a variety of ways, these standards make clear not only which forms of distortion are acceptable but also which are considered the least mediated and ultimately the most normative way of viewing the body. Harvey A. Zarem, the author of a chapter titled "Standards of Medical Photography" in Nelson and Krause's collection, argued that cosmetic surgeons have the same duty as laboratory scientists to make sure their data collection techniques (in this case photography) "are reliable and consistent."[15] Surgeons use these photographs as evidence of their skill, innovations, and efforts in advertisements for their practices, lawsuits, conference papers, and journal articles. While in the short term, pre- and postsurgical photos can help refresh the memory of surgeons, over a period of years, photography helps surgeons visualize larger "trends, evolutions, and outright discoveries" that can lead to more knowledge concerning plastic surgery and changes in the aesthetic evaluation of the human body.[16] These photographs also developed a "socioeconomic importance," as insurance companies rely on photographs when deciding whether to cover a patient's condition.[17]

Since cosmetic surgery has far less governmental regulation than most other medical fields, the industry collects photographs and other documentation as part of a medical record in order to standardize and legitimize its practices as scientific and safe. As late as 2012, many states did not require any accreditation or licensing of medical offices where cosmetic surgery operations take place; those that do often only regulate surgeries where the patient is fully anaesthetized or in the rare case when private insurance or Medicare covers the bill.[18] Along with surgeries that result in patients' being displeased by their postoperative appearance, this lack of regulation has led to cases in which patients died from infections after undergoing surgery in unsterilized environments performed by untrained doctors with tools and machines they did not know how to use.[19] While there are few statistics available on the number of lawsuits that result from these practices, in Great Britain, where many procedures are unregulated, cosmetic surgery made up "80% of the rising number of legal actions" in 2012, and 45 percent of these cases were upheld (compared to an average of 30% across medical fields).[20] Thus, cosmetic surgeons take and store photographs of patients for use as evidence in court to help curb such lawsuits when possible.[21]

Many of the authors of *Clinical Photography in Plastic Surgery* focused on how photographs are important not simply for surgeons but also for patients because they act as a "reinforcement to the patient that he or she is part of the process" and can lead to an overall greater "confidence in plastic surgery as well as in modern medical treatment."[22] In addition to allowing surgeons to learn from past surgeries and to protect themselves from lawsuits, photography is also regarded as a way for the patient to play—or at least feel as if he or she has played—an active part in the modification of their own body.

Digital Technology and Clinical Recommendations

The standards of beauty that were developed via photography are now used as a model to digitally transform physical images of potential cosmetic surgery patients. These manipulated images are then used as templates for actual physical changes through surgical operations. Just as plastic surgeons have long been trained in the proper practices of photographing the body, they now often use Adobe Photoshop and other more specialized programs focused on image (and virtual body) manipulation.[23] As new photographic and digital technologies have developed, articles on what the standards and practices should be have proliferated and appear often in widely read journals ranging from *Plastic and Reconstructive Surgery* to the *Annals of Plastic Surgery*.[24] These plastic surgery journals largely address the differences in the practices for recording, archiving, and presenting images created with celluloid film versus those created in digital formats. This creation of standards and best practices helps to make these digital images (with their more tenuous indexical status) suitable as evidence in both court cases and scientific studies.

At the same time, those cosmetic surgeons who offer such services often heavily advertise them on their websites as both cutting-edge and pleasurable techniques that separate them from competitors. Unlike the before-and-after photo, these digital manipulations not only guide the patient through the process of imagining what their body could eventually look like but also to help recommend particular bodies and manage the expectations and fantasies of patients concerning what kinds of bodies are possible. One article on the use of these technologies at the Parker Center for Plastic Surgery in New Jersey noted that "for any cosmetic surgery patient, trying to articulate exactly what they want done is often one of the most difficult parts of the entire process."[25] While these digital imaging programs can certainly help patients articulate their desired physical shape and try on a variety of new bodies with no consequences, these programs are primarily used to help the surgeon make a particular recommendation and convince the patient that that recommendation, whether or not it is what the patient originally came in asking for, is actually the best course of action. After patients discuss how they would like to look after the surgery and what parts of their bodies they would like to change, the doctor takes photos and digitally manipulates them to the patients' specifications—by changing the shape of the nose, enlarging the breasts, decreasing the size of the stomach, or making any of a number of other changes. Even though these programs can potentially be used to manipulate and morph these bodies in any imaginable way, in practice they are used to restrict possibilities by focusing only on those changes that are most normative. Thus, as digital technologies help manage new and often complicated and confusing personal choices concerning bodily modification, their recommendations guide patients toward particular surgical decisions that are often the most conservative, those focused on erasing difference and accentuating traditional, gendered beauty norms through the shape of the body.

Along with the relatively low-end two-dimensional programs like Anaface, there are many expensive three-dimensional digital imaging programs (over $40,000) that are specialized for cosmetic surgeons. Like many other companies in the field, Crisalix, the maker of e-Stetix 3D, a professional web-based cosmetic surgery simulation site that specializes in simulating breast implants, describes itself as aiming "to develop scientific technologies in the field of plastic and aesthetic surgery to improve surgeon-patient relationships."[26] e-Stetix 3D automates much of the process of body image editing and displays patients in three dimensions, allowing them to be viewed from any angle, naked or dressed in one of a wide variety of default clothing options. Crisalix specifically markets its e-Stetix 3D program by focusing on its ability to "increase consult-to-surgery conversions" by "boosting patient confidence."[27] On the Crisalix "testimonial" page, where one would usually see accounts from patients of their experience with the technology, Crisalix posts quotes from a number of surgeons, including Dr. Serge Le Huu, who raves that e-Stetix 3D increased his "conversation rate for breast augmentations from 60 to 86%."[28] Furthermore, Dr. Thomas Biggs, the editor in chief of the *International Confederation for Plastic, Reconstructive and Aesthetic Surgery Journal*, is quoted as saying that "the likelihood that the patient will make a positive decision regarding the surgery increases."[29] It is clear here that "a positive decision" can only mean surgery and, more specifically, a surgery recommended by Biggs and illustrated by the e-Stetix 3D program.

Moreover, these three-dimensional digital imaging programs are used not just to support a recommendation for an individual surgery but also to increase the number of surgeries per patient. Canfield's Vectra 3D camera and Sculptor 3D simulation software are part of one such high-end system that is sold to surgeons specifically as a way to "differentiate your practice" and "increase complementary procedures."[30] In a testimonial commercial for the Vectra 3D camera, a supposed patient lists all the different ways surgeons had tried to describe how she would look after breast surgery, including by showing her pictures of women with larger bust sizes and adding pads to her bra. She states, "No matter how hard I tried, I just couldn't envision myself with their bodies," but the Vectra 3D "could show how I'd actually look at different sizes." While she says, "I'd never imagined that realism like this was possible. . . . It gave me the opportunity to visualize what I was really going to look like," the image on the screen displays a torso with a bikini top that appears about as realistic as a Sims character.[31] This commercial illustrates how the discourses about these technologies still shape how we see ourselves through them. These images are no more real looking than those produced by any of the other methods for imagining your future body, even though that appears to be the technology's purpose.

This effort is alluded to on the website for the Beverly Hills Simoni Plastic Surgery Clinic, which explains the potential applications of three-dimensional imaging for patients who ask for rhinoplasty, since this operation "can potentially affect the balance of the entire face."[32] These images make it easier for Dr. Payman Simoni to argue that when changing the nose, other elements of the face might also need to be changed. His website says that "3-D imaging is one more way for him to communicate with his prospective Rhinoplasty patients what the goals for the surgery should be" so that he can lead them toward "sounder decisions."[33]

While these "sounder decisions" or recommendations for almost every body part appear differently depending on the particular patient, the altered nose always appears to be the same, with a concave curve and a thin bridge. The "goal" of rhinoplasty, according to these images, is always to achieve (or get as close as possible to) this particular shape, specifically because it is associated with whiteness and a lack of ethnicity. Many cosmetic surgery sites call these operations that seek to replace a "bulbous" appearance or "wide and flat bridge" ethnic rhinoplasties, as they seek to erase Semitic cultural markers from the patient's body.[34] This particular look is also one of the easiest to accomplish via most cosmetic surgery digital imaging software, as it requires just one or two movements of the mouse to erase the convex curve. By designing the software to make this technique so simple and by only showcasing this one particular shape of the nose that is specifically associated with whiteness, surgeons like Simoni end up severely limiting the types of bodies that patients can adopt and the forms of beauty that are encouraged to exist. Thus, these limitations help to suppress difference and celebrate the dominance of whiteness.

The logic of normalizing cosmetic surgery through guided digital imaging, wherein the surgeon will make as many changes as possible, reaches a high point on sites like virtualplasticsurgeryoffice.com, the now defunct online presence of a Los Angeles cosmetic surgery clinic that specializes in "extreme makeovers," which require multiple procedures that result in dramatically different appearances. This site supplies a great deal of information and encourages Skype interviews for potential out-of-town patients who might fly in for multiple surgeries at a time, including "designer laser vaginoplasty," "Brazilian Butt Lifts," and "Vaser Hi Def Liposculpturing." When getting "a breast job, bootylicious J-Lo butt, and a Designer Vagina all at one time," this clinic offers its trademarked "Wonder Woman Makeover," also commonly known as a "Mommy Makeover."[35] This package of surgeries, which often costs between $10,000 and $15,000, seeks to eliminate all signs of pregnancy from the body of a new mother. In doing so, as Natasha Singer of the *New York Times* phrases it, it "seeks to pathologize the postpartum body" and treats "motherhood as a stigma."[36] Singer argues that, by packaging these surgeries together, the cosmetic surgery industry turns "badges of motherhood" into "badges of shame" in order to create a new and very lucrative market to treat a pathology that previously did not exist.[37] Yet, while cosmetic surgeons may profit from the pathologizing of motherhood, it is a bit of a stretch to argue that they are solely responsible for it. At the same time, the normalization of this surgery also makes it clear how normal and common these all but entirely hidden postpartum bodily transformations are and how connected this new body is to the fantasies, pleasures, pains, and realities of child-rearing. By packaging and recommending multiple surgeries at once, clinics increase the likelihood that these mothers will get them all, even if they originally only wanted one. They also exacerbate the problem that the markers of a mother's body are perceived as a stigma best hidden or smoothed away altogether.

Cosmetic surgeons thus use digital imaging as a tool to accentuate patients' visions of themselves as endlessly flawed but also correctable for a price. Theorist Susan Bordo describes an experience that occurred when she went to a dentist to repair a discolored tooth. Her dentist created an image of her post-operation mouth and also recommended a handful of other, much more invasive operations that would improve her overall smile. Although she told him she was not interested, he still mailed her "a computer-generated set of recommendations" written in such a way as to indicate that "all the options that were *his* ideas" actually represented her own "expressed dissatisfactions and desires."[38] These new recommended operations that she ultimately turned down consisted of much more invasive procedures on her gums and had a price tag of $25,000. As Bordo argues, every mark on the body becomes magnified by these technologies, as they tend to teach subjects to view their bodies as being ever more acutely visible.[39] This way of looking at the body commodifies every portion of it and trains subjects to view it as a "fixer-upper," full of defects that must be worked on and eventually replaced. Repudiating Davis's argument that most cosmetic surgeries are performed on patients who just want to appear "ordinary" rather than beautiful, Bordo argues that "in a culture that proliferates defects and in which the surgically perfected body ('perfect' according to certain standards, of course) has become the model of the 'normal,' even the ordinary body becomes the defective body."[40]

Most, if not all, surgeons make sure to specify that digital imaging is not a guarantee of outcome, often in written legal documents and disclaimers, if they are going to offer such services. While these technologies can get plastic surgeons into legal trouble and can only serve as a sketch rather than a schematic during surgery, they are still very popular because they add to the plastic surgery experience the ability for the patient to experiment with their own body and reshape and reenvision it in ways that her or she might not have had a chance to before. Oddly, this technology allows for an excitement and enjoyment of cosmetic surgery that the surgery itself does not actually produce—or at least that is what these technologies' marketing departments would have prospective patients believe. Recommendations and imaging here work to turn surgery into a commercial enterprise and the patient's body into a commodity while they simultaneously make patients feel as if they have agency in the process.

Body Morphing from Home

While the use of body-morphing digital imaging technologies started in cosmetic surgery clinics as a way to guide the personal choices of patients and homogenize the kinds of bodies and beauty that are deemed acceptable, these technologies have now become part of the larger digital media environment.

Just as they became popular for patients because they are enjoyable to use and turn the body into a site of possibility and potentiality, they are now popular as a source of online entertainment. By focusing on the possibilities of bodies, cosmetic surgery as an industry has achieved its original goal of using photography (and the surgical recommendations that sprang from it) as a way to not only standardize and legitimize itself but also become a normal, if not everyday, activity around which a variety of online communities have formed.

These digital imaging programs entered online spaces first via companies that allow both potential patients and casual users to email (or mail) in pictures of themselves with instructions concerning what types of surgeries they would like to simulate. For anywhere between $19.99 and $69.99, these sites send back a modified image to prospective patients of what they would look like if such changes were implemented. The Beauty Surge, "the cosmetics surgery supersite" that acts as a community center, resource, and store for those interested in discussing or getting surgery, became an early purveyor of this service in 2004 when it announced that, through "advanced software and an in-depth knowledge of plastic surgery," it would now offer patients the chance to see the "expected results" of their proposed surgeries.[41] Without any real conversation, editors will digitally transform users' photos in whatever way the editors choose. This removes patient agency even more so than occurs in clinics, as the patients have no chance to take control of the interface or interject with their own ideas of how they would like to look.

Other sites assert that the purpose of these digital manipulations should only be testing out new bodies, transforming one's image rather than one's actual body via surgery. Foto Finish advertises its "digital plastic surgery" service by explaining, "Now there is no need to go under the knife . . . at least as far as your pictures are concerned. Foto Finish artists can take years off your face and neck, remove scars and blemishes, whiten teeth, remove lipstick from teeth, and correct other unsightly problems. We can digitally apply makeup, fix stray hairs, and more. You'll still look like you . . . only better!"[42] These sites stress that any kind of manipulation is possible in a photo and also propose that image manipulation is a useful and acceptable practice, especially when the changes you want to make are somewhat minimal and would not necessarily be noticed by someone meeting you in real life for the first time. For the purposes of presenting yourself to acquaintances on Facebook, getting a job via LinkedIn, or even getting a date on an online dating site, the possibility of manipulating your photo is advertised as a subtle way to "improve" yourself. The changes in these photos must be subtle so that those people that users meet via these sites assume that they had an excellent but real photo rather than a digitally altered one. However, since so much of one's life is now conducted online, one's digital image has become as important as—and in some cases more important than—one's actual appearance. At the same time, such technologies may in fact encourage some users to opt for actual surgery eventually.

As in other domains of recommendations and self-management, self-help, or self-care, transforming the self is here presented as a form of pleasurable agency associated with autonomy, freedom, and physical transformation. Along with simply paying someone to change your image, many online tutorials now teach how to do it yourself via Photoshop. One suggests that the process is "so simple and easy, you'll be tempted to use it on photos of people who honestly don't need it."[43] By asking the user to view everyone as a possible candidate for rhinoplasty, regardless of their need, simply because it is "so much fun and so easy to do!" this web tutorial encourages a surgical gaze that is common throughout websites devoted to digital cosmetic surgeries.[44] These sites ask users to view and modify everything as if they were magazine editors who feel the need to transform the shape and dimensions of their cover models.

Rather than becoming a site for genuine free play, however, the digitized body becomes a site of further discipline. One website notes that "digital technology can liposuck you, hide every fault from bumps to lumps, wrinkles to crinkles, complexion, hair, cellulite and teeth which are all digitally enhanced and every imperfection is fixed to achieve a flawless look."[45] Through this focus on the photographed and transformed body, this surgical gaze homogenizes not just beauty but, as Bordo puts it, normality through the figure of the digitized body.[46]

Moreover, recommendations in the form of online play with the digitized body are often intended as a gateway to actual body modification. Indeed, cosmetic surgeons have been very successful in normalizing and popularizing their profession through the dissemination of online digital imaging programs designed to both ease potential patients into the clinic and expand their customer base to a wider clientele. For example, on FaceTouchUp, for $9.95, users can upload up to ten pictures of themselves and use the site's simple tool to edit their bodies in the way they would like. This website specifically targets users who are interested in cosmetic surgery but are "on the fence."[47] After users modify their images, the site encourages them to download the images or email them directly to their surgeon in an attempt to make these currently uncertain users into full-fledged patients. While doing this at home by themselves certainly allows users to experiment with their bodily shapes in ways that they would not be able to in a surgeon's office, there are still many guides on the site that encourage them to shape their bodies in the same ways they would be shaped if a surgeon were making the recommended edits.

Before modifying their own bodies, users are encouraged to practice on one of three preloaded faces, a white woman looking to the right, a white man looking to the right, or an African American woman looking straight ahead. Since this area is free and the controls allow for a wide variety of physical transformations, it can be more readily used as a space for imaginative play and pleasure. Yet, instead of being focused on playing by transforming or distorting your own body, this body manipulation is focused on others and can become an exercise in mastery and control tinged with racism and sexism instead of self-expression. While the angle on the two white people showcases the slightly convex shape of their noses, the African American woman's photo foregrounds the broadness of her nose. This program showcases non-Caucasian ethnicity (both Semitic and African) and recommends that users transform or delete it. Video tutorials on the site first explain how to transform these noses to make them narrow and concave (i.e., less ethnic). One tutorial goes further and proposes that after completing one edit, the user should attempt "a more ambitious procedure" and transform other parts of the face to show how the individual would look with not just rhinoplasty but also a chin augmentation and neck lift.[48] Just as in a surgeon's office, this digital imaging tool encourages users to transform bodies to align with a conservative and specifically white ideal of beauty through as many costly surgeries as possible. During this process, users are taught to view bodies from the perspective of a cosmetic surgeon.

Along with the tools on its website, FaceTouchUp sells desktop digital imaging programs to surgeons for use in their practices and will also create custom iPhone and iPad apps to help advertise their clinics. The fifteen apps that the company has so far made, which are all advertised on its site, all look very similar and all include the same basic services, such as videos, contact information, and the same morphing technology available on the website. After users have modified their images, they are encouraged through an automatic prompt to email the surgeon for a price quote, send the image to a friend, and publish it on Facebook.[49] Through these prompts, the software pushes users not only to become more serious about being a patient but also to make this public knowledge and a topic of conversation through social networking. These tactics are a way to transform cosmetic surgery from simply a one-time (and often hidden) event into a normative lifestyle choice around which communities can form.[50]

This effort to create communities and lifestyles around plastic surgery is a common thread that runs throughout many online digital imaging programs. For instance, Crisalix markets its three-dimensional imaging technology to cosmetic surgeons via Crisalix.com and also hosted the recently defunct Sublimma.com, a website that marketed e-Stetix 3D technology and breast augmentation surgery specifically to patients and consumers. With a presence on Twitter, YouTube, and Facebook—where roughly 1,400 people have "liked" it—Sublimma still offers a variety of ways to create a community around a shared interest in breast augmentation. While Crisalix offers testimonials from surgeons who valorize the software's ability to convince patients to get a surgery, Sublimma offered quotes from patients who largely describe the way the technology taught them a new way to view themselves. One person remarked, "Seeing myself in 3D allowed me to look at my body from a whole new perspective," as if this 3D effect were more revelatory than viewing oneself in a mirror or on video.[51] Many other patients remarked that their surgeons would recommend a particular implant for them and then use e-Stetix 3D to visualize their new bodies, which they described as "a real motivational factor!"[52] By focusing on simply visualizing oneself in a different body (rather than visualizing one's body differently), e-Stetix 3D works to construct cosmetic surgery as an act of self-pleasure or narcissistic self-care.

Although Sublimma and FaceTouchUp have discussed digital imaging primarily as an educational (or disciplinary) tool that can help users see their bodies from "a whole new perspective," NKP Medical Marketing's SurgeryMorph is much more interested in the entertainment and playful aspects of the tools. While David Phillips, the CEO of NKP Medical Marketing, commented on these tools' ability to take "the guesswork out of the process," the reason he believes it is really special is that it does so in "a manner that is both fun and easy."[53] On the SurgeryMorph website, this hypothetical surgery technology is described at one point as "an exciting and fun tool."[54] A connected video tutorial simulates liposuction, blemish removal, and breast augmentation using a photograph of an attractive white woman in a bikini while the Beach Boys' "California Girls" plays as a soundtrack. Users are then encouraged to share their modified images on Facebook, where they will appear with the statement, "I have been using NKP Medical SurgeryMorph to play with different cosmetic procedures! NKP Medical SurgeryMorph is available at www.surgerymorph .com."[55] Since, as the video exclaims, this technology is "fun-easy-viral," even if people use the software without actually getting one or multiple surgeries, their Facebook friends may see the pictures and decide to try this virtual service out themselves. By encouraging users to make their virtual surgical fantasies and interests public, SurgeryMorph changes cosmetic surgery from a hidden and shameful act to a celebrated act of personal agency and fun consumption; it becomes a process driven by recommendations. This focus on sharing modified images via Facebook, Twitter, and other social networking sites is common to many technologies, ranging from BodyPlastika's iPhone app to Plastic Surgery Before and After's downloadable software.

Indeed, these digital imaging technologies and sites have proliferated and have themselves become standardized. All of these programs use the same basic tools and look quite similar. While the programs have become standardized, they have also been adopted for other kinds of beauty and bodily enhancement, ranging from skin-care and makeup visualizations to programs that allow you to try on clothing on your iPhone. For example, ModiFace, a company started by Parham Aarabi using research he completed in graduate school and as a tenured professor and Canada research chair at the University of Toronto (positions he received at the age of twenty-four), works to create and improve "face processing algorithms," the kind of underlying code that is the basis for all cosmetic surgery digital imaging programs. Aarabi's work won a worldwide innovation award from MIT and, according to the ModiFace website, he and his company have published 112 scientific papers on the topic.[56] With this technology, ModiFace specializes in creating virtual makeover applications for major media outlets, magazines, and cosmetic brands, including a wide range of clients like Garnier, Stila, Good Housekeeping, Harpers Bazaar, and NBC Universal. Its original algorithms and software, used first in digital imaging programs that specialize in cosmetic surgery, are now used as the basis for a huge variety of makeover tools that are central to over sixty different online and computer applications and thirty different iPhone apps. Over five million of its apps have been downloaded and they have a million active users.[57] If one uses a makeover application on the website of a major media or cosmetics outlet, whether it be to test out an antiaging cream, a new lipstick color, a wedding gown, or a face lift, he or she is more than likely interacting with ModiFace.

While most other digital imaging programs require the user to reshape their faces by pushing and pulling on the skin manually in the image, ModiFace detects different facial areas (eyes, nose, mouth, etc.) and performs these edits automatically. On LiftMagic, a ModiFace "instant cosmetic surgery and antiaging makeover tool," users can upload their images and have the program simulate up to sixteen different surgical operations at once, ranging from various "injectables" to rhinoplasties and neck lifts. After LiftMagic automatically generates a postsurgery image, it offers to upload that image to one of several other ModiFace websites, where users can digitally lose up to fifty pounds of weight or try on makeup and "1000s of celebrity hairstyles."[58] After users have completed their extreme digital makeover, they are offered the possibility of seeing how their image would look on a billboard in a virtual but photorealistic Times Square or in a number of other public venues using PicMee.com. In the process, these virtual body modifications and recommendations quickly become first normalized and then spectacularized. While ModiFace and many other digital imaging programs are designed for private use and personal bodily experimentation, by offering people the chance to share their images on Facebook or the side of a skyscraper, such sites encourage users to think of their bodies as a spectacle and their actions as part of a larger culture of celebrity and homogenized beauty.

Although ModiFace's core business focuses on supplying cosmetics companies and surgeons with facial imaging programs, it also makes several free apps that showcase its programs to those who may be more interested in simply experimenting with their bodies and less interested in any actual transformations. For instance, its Makeover app allows you to change your eye color and hair and see what you might look like wearing different types of makeup and jewelry. After modifying your image, the program will place it on the cover of a fake fashion magazine and list all the makeup you would need to buy to look the way you do in the image. This is the equivalent of kids playing dress-up or experimenting with makeup. For an extra price, you can also add the ability to change the dimensions of your face to see what you might look like with a face-lift or Botox. On the one hand, programs like this can be a good way of learning more about your body and becoming familiar with how you actually look in a playful rather than judgmental way. On the other hand, in this setting, playing dress-up is a step on the way to more intense physical transformations that are fraught with aesthetic bodily judgments. The resulting images are not what one might call natural, but they do capture what one might look like plastered with makeup and wearing a wig. Other ModiFace programs, including ModiFace Live, will also make your eyes extra large like an anime character's or add odd, Snapchat-like effects to your image such as cat eyes or sparkles. The software is also clearly designed only for women; for me, the program is most useful as a way to see myself as a passable drag queen.

Capitalizing on this trend, Dr. Michael Salzhauer at the Bal Harbour Plastic Surgery Clinic in Miami released his own digital imaging software both as a feature on his webpage and as iSurgeon, an iPhone app. While still an advertisement for his clinic and for plastic surgery more generally, this app has a very playful tone and uses fonts that appear to be inspired by the look of bubble-gum. While users manipulate their faces, the sounds of drills and other surgical devices play in the background, which is both jolting and ridiculous. During one television news interview, Salzhauer took a photo of his interviewer and used iSurgeon to make her look more like a witch (with a long nose and a recessed chin) in order to show off that, "in the end, this app is for fun and is meant to help you satisfy your curiosity about how you would look with that elusive perfect body and face . . . or not."[59] MTV found the playful tone of this app interesting and built it into its Heidi Yourself program, in which users can have a chance to see themselves after "surgery like the stars."[60] Named after Heidi Montag, a star of MTV's The Hills reality television series who reportedly had ten cosmetic surgery operations in one day, this website parodically offers users the chance to see how so many surgeries would transform their own bodies. Even as Montag herself has expressed regret concerning her extreme makeover, this site encourages users to experience the same effects. As one blogger put it, "You can do some interesting things to yourself with those tools and by the end, I look pretty crazy."[61] Yet, even though Heidi Yourself includes the sounds of screeching drills, it is hard to view these surgeries as a cautionary tale; instead, the whole situation simply appears absurd. While one may not want to do anything as drastic as what Montag did, this program does not judge her for her choice and instead encourages users to exercise their own personal choice in deciding whether to digitally (and then surgically) transform their bodies however they like. Programs like this work to blur (if not entirely erase) the line between the empowering and pleasurable experiences of physical experimentation and the dominating capitalistic and commodifying logic of the cosmetics industries that regulate which forms of experimentation are body positive and which are objectionable.

Collaboration and Beauty Rankings

Sites that help users experiment with transforming their own bodies also often encourage them to invite others to weigh in on the decision through Facebook messages and other collaborative social means. Friends on Facebook are asked whether these possible surgeries are a disaster or are actually flattering and worth the danger. While these outside perspectives from friends can act as a voice of reason, they also impose a normative ideal of beauty; this virtual social environment and the recommendations that come out of it affect how users make their personal choices concerning bodily modifications.

While not programmed into code, this method of bodily recommendation is primarily collaborative in nature, but unlike the other collaborative filters I have discussed throughout this book, this method of collaboration usually creates recommendations that are only based on the opinion of a few people, rather than millions. While the use of Facebook to both advertise and get advice on cosmetic surgery is common, there are also many sites that take a more direct approach to using collaboration to generate recommendations. On Make Me Heal, a site that sells skin care products and cosmetic surgery supplies ranging from Botox injections to post-op mastectomy bras, users can upload images of themselves and ask other users both whether they should get surgery and what kind of surgery to get. While few people comment on these pictures, many of them receive thousands of views and votes on whether they really need the surgeries. Users also post images of their completed surgeries on this site, whether successful or not, and provide insight to the community on how to deal with any problems that might arise. Through this recommendation apparatus, users form communities on the site and become acclimated to a surgical culture.

These exercises in judging which types of bodies need surgery and which do not are also built into the Virtual Plastic Surgery Simulator. This digital imaging program, available for use on personal computers, a large variety of mobile devices, and as a Facebook app, allows users to manipulate their photos through blurring, stretching, and warping.[62] After completing a modification, users are given the option of uploading their images to the company's Befaft.com website, where other users can vote on whether the proposed surgeries should actually be done. These other users are asked to decide both whether the proposed changes are actually an improvement and whether this improvement is large enough to make it worth the cost and pain of the necessary surgeries. This process turns the program into an entertaining game for the user base as a whole, as they can see how many people think that particular surgeries should be done and how many people think they should not. Often, the community does not agree or disagree, and users can also write responses and explain their reasoning for why a surgery should or should not be done. If users allow it, Befaft.com will then automatically publicize their pictures, votes, and comments by posting them on the users' Facebook and Twitter accounts. By sharing these images, the program teaches users how the community as a whole defines beauty, and what kinds of appearances one should strive for when engaging in both digital editing and surgery. Like many other programs, this one also offers tutorials on how to use the virtual surgery tools on its site and demonstrates the proper ways to edit your body. Along with demoing breast augmentation, weight loss, liposuction, and other typical surgeries, this site also displays the results of what it calls an "ethnic" surgery, depicted in before-and-after pictures of a dark-skinned woman who has elected to get rhinoplasty to narrow the bridge of her nose.

This method of recommendation is heavily influenced not just by the culture of plastic surgery but also by the collaborative filtering of social networks. Make Me Heal and Befaft.com are both aesthetically and functionally very similar to Hot or Not, a site where people can rate others based on photos. This site, which itself is based on FaceMash, Mark Zuckerberg's first foray into social networking, wherein people would vote on which Harvard students were most attractive, is now not simply a site to judge beauty but also a site where people are encouraged to get in touch with others who are voted to be of a similar level of attractiveness.[63] Friendships, romantic relationships, and whole communities have grown out of Hot or Not that are largely based on similar bodies rather than interests. Like Hot or Not, digital imaging sites generally create communities by recommending ideals of beauty and the surgeries necessary to attain them.

While these sites most commonly rely on actual users to collaboratively recommend whether to get a surgery, there are efforts to automate this process of voting and even to demonstrate specific surgeries to potential patients. The particular issues concerning how algorithmic recommendation technologies can affect how the body is imagined are brought into focus by the "beautification engine," an experimental software for rating and transforming images of the human face. Developed by Tommer Leyvand at Tel Aviv University and Microsoft and presented at SIGGRAPH 2008, the annual meeting of the Association for Computer Machinery's Special Interest Group on Computer Graphics and Interactive Techniques, the beautification engine "uses a mathematical formula to alter" an image of a face "into a theoretically more attractive version, while maintaining what programmers call an 'unmistakable similarity' to the original."[64] Leyvand programmed this engine with a set of collaborative filtering algorithms similar to those used by TiVo and Netflix that try to rep-licate the recommendations that communities on sites like Make Me Heal would offer. However, instead of measuring interest and desire, the beautifica-tion engine measures 234 different parts of a face, including such things as the distances between one's eyes and between one's lips and chin.[65] The program also contains a list of the sets of measurements that are supposedly the most ideal and the same types of measurements used by surgeons themselves. The beautification engine works by looking through the preprogrammed measure-ments to find the set that is closest to the image of the user's actual face. It then tries to transform the user's image to make it match up as closely as possible to an ideal set of measurements while still remaining recognizable as the user, leav-ing "the person's essence and character largely intact."[66] Leyvand imagined that such digital manipulations could become a common procedure in com-mercial photography's "ever-growing arsenal of image enhancement and retouching tools available in today's digital image editing packages. The poten-tial of such a tool for motion picture special effects, advertising, and dating services, is also quite obvious."[67]

In many ways, this technology embodies a Pygmalionesque impulse to imprint certain dominant cultural modes onto another's physicality. A *New York Times* article on this technology mentions that the researchers created their program specifically with a "beauty estimator" algorithm that focused on white notions of beauty. Their beauty estimator was in fact based only on the responses of sixty-eight men and women, aged twenty-five to forty, who were all born in either Israel or Germany. This tiny subset was then shown "photo-graphs of white male and female faces and picked the most attractive ones."[68] Leyvand and his coauthors argue that while this subset is small, the resulting beautification engine had universal appeal and could transcend "the boundaries between different cultures, since there is a high cross-cultural agreement in facial attractiveness ratings among different ethnicities, socio-economic classes, ages, and gender."[69] Leyvand has also stated that "attractiveness ratings are, in fact, universal." He argues, "Beauty is not in the eye of the beholder. . . . If I took the same photo and showed it to people from 10 different regions with 10 dif-ferent backgrounds, I would get roughly the same results."[70] Even if very diverse, a group of sixty-eight men and women is far too small to base such a universal conclusion on. He further argues for the universality of his findings by point-ing out that even infants spend more time looking at attractive faces, "regard-less of the faces' gender, race, or age."[71]

Similarly, in 2016, Wayco Enterprises in Hong Kong, a manufacturer of beauty products among other things, created Beauty.AI, "the first international beauty contest judged by Artificial Intelligence." Thousands of people submit-ted photos to be judged by algorithms. Of the forty-four winners announced, only one had dark skin and very few were Asian. The creators of the algorithms explained that the "mistake" was caused by their algorithms' being trained predominantly on light-skinned models, which created a bias in their judgments.[72] Law professor Bernard Harcourt stated that "the idea that you could come up with a culturally neutral, racially neutral conception of beauty is simply mind-boggling" and compared this example to the many ways racism and sexism enter into all forms of algorithmic prediction from predictive policing to Twitter bots.[73]

Nevertheless, the creators of the Beauty.AI contest stated that in their next attempt they would try to eliminate discrimination in their program.[74] Similarly, while Leyvand and his research group have yet to make a beautification engine specifically for "nonwhite racial and ethnic groups," they reported that they would work next on trying to change their algorithms to modify not just the geometry of the face but also the color and texture of hair and skin.[75] It is more than a bit ironic that people from the two countries that are most closely identified with the horrors of genocide and eugenics are now responsible for coming up with a new standard of universal beauty that is easily compared to the one created by Third Reich. Indeed, Leyvand received much of his inspiration from Thomas Galton, the father of eugenics, whose work, as I argued in chapter 4, also influences online dating site recommendations.[76] Galton made composite images of various groups of people, including convicted violent criminals, by superimposing their images on top of each other.[77] He would then measure the distances between the facial features on these composite images in order to divine what an average or ideal criminal, vegetarian, or diseased person would look like. He also imagined that his invention would be most useful for identifying an average member of a race or a family (by superimposing images of siblings) or even producing an ideal image of a specific person (by superimposing their image from various periods).

This method for empirically defining beauty remains prevalent, and Leyvand relies on the same basic phrenological concept in his efforts to define beauty and mold people to this image.[78] Importantly, what is noticeable in the transformed images is that "a blandness can set in."[79] Those "imperfections" that are often the most striking aspect about a person are frequently stripped away by these algorithms, as they seek to make everyone "more attractive," specifically by making them less physically diverse. Galton's theories have recently regained traction. As the authors in the edited collection *Facial Attractiveness: Evolutionary, Cognitive, and Social Perspectives* make clear, many now challenge the idea that beauty is in the eye of the beholder via both scientific and cultural studies. Overall, this collection argues that while "certain extremes, symmetry, youthfulness, pleasant expressions, and familiarity all contribute," "averageness is the key to attractiveness."[80] By automating the process of evaluating beauty and by having physical alterations take place in a digital photo rather than on a person's actual body, Leyvand's beautification engine reveals how radically the relationship between users and their physical representations is changing at a moment when it is often far more likely to be seen and recognized online than in person. This algorithmic standardization of beauty becomes the basis for the recommendation of new bodies. As psychologist Nancy Etcoff explained, "Everyone wants to look better. And we keep taking it further and further to all these images that have been doctored. There is a whole generation of girls growing up who think it's normal not to look the way they really look."[81] This trend is only growing as social media websites like Facebook and various dating websites become more important for communication, and it is less and less important to have images of a user exactly represent their actual appearance.

While the beautification engine tries to automate digital imaging and the rating of beauty, there are many other programs that try to accomplish this by relying on proprietary algorithms that claim to be objective and accurate because they are based on both classic, "timeless" ideas of beauty and contemporary scientific research. These ideals importantly always privilege whiteness and make ethnicity a problem. Many of these programs and sites, including Anaface, ModiFace, the Beauty Analyzer iPhone app, and FaceBeautyRank .com, refer to various classical ratios, such as the golden ratio or the Fibonacci sequence, as the numerical (if not numerological) foundation that underlies the way we compute aesthetics. Developed by Euclid around 300 B.C., the golden ratio (represented in algebra as phi) is a geometric ratio apparent throughout nature in the shape of flower petals, pinecones, shells, galaxies, and DNA molecules. In terms of physical beauty, many have used the golden ratio to define an ideal face as "two-thirds as wide as it is tall, with a nose no longer than the distance between the eyes."[82] In 1935 Hollywood makeup artist Max Factor used this ratio to create a "beauty micrometer" device that used metal strips and screws to help other makeup artists reveal minute defects in actors' faces and register their "facial measurements and disclose which features should be reduced or enhanced in the makeup process."[83] Resembling Pinhead from the Hellraiser franchise, this device appears to be an instantiation of the most disciplinary and imprisoning aspects of makeup and beauty culture. At the same time, it importantly also was only ever meant to be used on people being filmed, as close-ups might reveal details otherwise invisible to the naked eye and the makeup fixes for these "flaws" would be far more noticeable than the flaws themselves. In the process, the micrometer illustrates an early effort to focus more on how individuals look through technology than in real life.

In an effort to update this work, cosmetic surgeon Stephen Marquardt in 1997 created a new mask and three-dimensional digital template that could be applied to people's real or virtual faces to locate flaws. He developed this mask specifically to guide users toward surgical procedures to make their faces "more like the mask" and therefore more attractive.[84] However, he also envisioned a

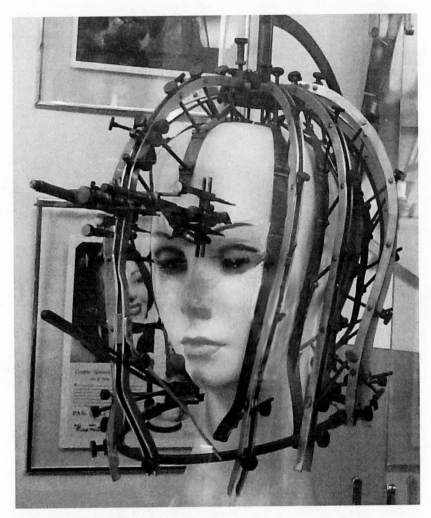

FIG. 5.2 Max Factor's Beauty Micrometer

time when people could modify not just their faces but also their DNA to better fit his aesthetic model.[85] Rather than celebrate the beauty of difference in all its forms, Leyvand, Factor, and Marquardt all propose the positive benefits of the homogenization of physical identity and the principle that through physical and genetic conformity, everyone will become beautiful.

Like Anaface, Face Beauty Rank and the personal computer software the site sells allow users to "evaluate the beauty of a face" based on the golden ratio, which it refers to as "geometric aspects of beauty" that are "somewhat 'stable' as compared to opinions and views."[86] While other sites argue that such evaluations should be performed for their own sake, Face Beauty Rank offers its services primarily to people who are trying to find an objective, if seemingly random, way both to pick which of their photos to post on dating websites and to choose which other users to contact. Its site advertises, "Let Face Beauty Rank help you decide on your new date! Rank human faces objectively and automatically with a computer program to help you make a decision."[87] At the same time, underlining the horoscope-like nature of these recommendations, the website contradictorily states at the bottom that it "cannot be a trustworthy tool for assessing beauty" and asks users to "not take the results of the program too seriously!" It cautions, "Do not criticize your self-esteem and the self-esteem of other people judging from the Beauty Rank!"[88] Yet it also markets its software directly to those developing dating websites as a way to automatically select the best pictures of users, which will then help "visitors find their perfect match!"[89] Face Beauty Rank sells itself as a program to help people better present themselves online, even if they have trouble making basic choices concerning how they want to appear.

The no longer available Beauty Analyzer iPhone and Android app also rates users' faces based on the golden ratio. Created by the Nascio Development Company, the Beauty Analyzer automatically rates the faces of users on a one-hundred-point scale. It also gives users detailed feedback on their various facial features and how well proportioned they are in relation to one another. Nascio advertises that while it may be fun to use, it is a serious app that provides "genuine results based on scientifically tested rules."[90] The algorithm behind this app is based on a 2007 academic paper by Kendra Schmid, David Marx, and Ashok Samal that reports the results of a study that sought to "predict the attractiveness of a face using its geometry" and its relationship to "neoclassical canons, symmetry, and golden ratios."[91] This study asked thirty-six students and employees at the University of Nebraska to examine 232 photos featuring random Caucasian people and Hollywood stars known for being attractive and rate the subjects' attractiveness. From this information, the researchers created a list of what the ideal facial measurements were and noted that the golden ratio and other related neoclassical mathematical conceptions of beauty do have "an important role in attractiveness."[92] As with the beautification engine and the Beauty.AI contest, given the study's small and very localized group of participants and its focus only on Caucasians, these findings skew definitions of beauty in ways that favor those with lighter skin. Rather than consider these effects, the article that reports the study results concludes by asserting that human perception of attractiveness is ahistorical and biological and can be objectively mapped and rated.

The golden ratio and the mathematical basis for beauty have become such engrained ideas that many math tutorial websites now use images of beautiful people to teach the principles of ratios and fractions. One such website showcases a variety of images that feature the golden ratio and natural elements that approach the golden ratio, including Leonardo Da Vinci's *Vitruvian Man*, the Greek Parthenon, and Jessica Simpson's face, to show how "beauty is mathematical!"[93] Users are then urged to measure the faces of various celebrities to see if they are really as beautiful as we think they are. These automatic recommendation technologies, beauty-analyzing apps, and other tools that rely on the golden ratio and other ancient Western aesthetic ideals proliferate the idea that beauty is an objectively measurable trait and promote the homogenization of human bodies as a fun and playful exercise.

These playful technologies that rate and recommend beauty have also been used to modify bodies to imagine how they will look years later. While certainly not commonplace, technologies of this sort have begun appearing in expensive self-management equipment that keeps track of aging baby boomers' daily regimens, including their levels of exercise, smoking, eating, tanning, and drinking. One of the first of these products, which was designed to be sold to medical clinics (though never actually released), was Accenture's Persuasive Mirror, an in-home system designed to help users "maintain a healthy life style."[94] This technology includes various cameras and a mirror surface designed to also be used as a bathroom mirror. A camera in the mirror itself takes pictures of users and generates an image of what they will look like in the future based on their current lifestyle. The Persuasive Mirror also relies on surveillance cameras placed throughout a person's house to gather information about the user and their lifestyle. The image that the "mirror" produces changes based on the users' daily activities, making them look better based on certain behaviors and worse based on others. In the spirit of Charles Dickens's Ghost of Christmas Future, this generated image is meant to show the user the future consequences of current behaviors in order to recommend new lifestyles. The mirror adds weight, wrinkles, and other signs of aging to the bodies of users who have not been exercising or sticking to their diets. To further encourage change, the transformed image also tries to incorporate colors and shapes that the surveillance equipment has discovered that the user dislikes. Thus, this future body is made ugly. In contrast, if the user has been "good," he or she will be greeted with an image that tries to portray the user in the way that the surveillance equipment gathers the user would like to be viewed. Thus, this technology visualizes different futures to motivate users to achieve long-term goals.

This software, with its focus on trying to influence the everyday actions of users in order to educate and promote everything from healthier eating habits to world peace, is part of a larger movement in the computational sciences known as captology, an acronym for "computers as persuasive technologies."[95] Coined by Stanford professor B. J. Fogg in 1996 and discussed in detail in his 2003 book, *Persuasive Technology: Using Computers to Change What We Think and Do*, captology focuses on the question of how software designers could use "the power of computers to change what people believed, how they behaved,"

FIG. 5.3 Image from Accenture Persuasive Mirror press release, 2006

and how this could then "be applied to improve the world in some way."[96] Digital recommendation systems, with their ability to guide users toward "correct" choices, are central to this way of employing computers. Fogg's description is a utopian version of Gilles Deleuze's control society made possible by the domestic surveillance networks of programs like Accenture's Persuasive Mirror. Like a life coach, such products control us by keeping track of all our actions and heavily recommending certain actions over others by making all other actions literally appear monstrous. While at one point extremely controversial because of fears about the kind of society that this form of control would create, these persuasive technologies are now central to the internet through, most notably, recommendation systems. Just as technologies ranging from Apple's operating systems to Gmail and Facebook try to transform to better fit each individual user, they also work to reshape the user to better fit the technology.

As I have argued throughout this book, recommendation systems and other captological technologies play a vital role in facilitating users' subjective understanding of themselves so that they better fit into a postfeminist neoliberal culture. Yet, when applied to human bodies, captology is concerned not simply with influencing human beliefs but also with shaping what it means to be human. While these technologies are used to actively reshape the contours and dimensions of the human body, many of those that currently exist are focused on reshaping them in order to help sustain dominant ideologies about beauty and aging.

While digital imaging technologies, beauty rankers, cosmetic surgery recommenders, and self-management technologies can be used simply to learn about and find pleasure in one's appearance, they also stress that the possibility of having a young and tight-skinned body is based not in biology but rather in personal choice. The Persuasive Mirror asserts that having an ideal body is an achievable goal and all one has to do to attain it is make the choice to transform one's lifestyle. By stressing that the shape of one's body is the result of continual life choices, the Persuasive Mirror asserts that those who cannot attain a Vitruvian body or a perfectly balanced face are simply not trying hard enough and suffer from a lack of both self-control and self-esteem. Yet the option to become fitter, happier, more beautiful, and therefore more economically valuable is one that is much more available to those who can afford it. Technologies like the Persuasive Mirror, with its complicated surveillance equipment, if ever released, will be extremely expensive and available only to the most affluent. Beyond this, while captological technologies identify a causal link between an attractive physique and happiness, cosmetic surgery makes it unnecessary to change one's lifestyle to attain this new body.

The use of photography and digital imaging in and out of plastic surgery clinics encourages people to view their bodies as plastic and modifiable within a consumer marketplace. Whole communities have been created by these technologies and this shared practice of looking. This activity of regulating, recommending, and automating these bodily modifications in a digital space has played a fundamental role in situating the body as a contested site of personal choice. Such technologies teach us to view ourselves from their perspective, which all too often is defined by a few white reviewers judging small data sets full of white models. In the process, they both show us what we look like from an algorithm's perspective and encourage us to shape our bodies, subjectivities, and culture in their image.

Conclusion

• •

On Handling Toddlers and Structuring the Limits of Knowledge

For parents dealing with toddlers who are learning "the ability to express their wishes," through both new vocabulary and tantrums, parenting manuals often recommend offering choices as a way to guide behavior: "Unfortunately, their wishes often differ from our own. So how do you allow your children to express their desires—and get them to cooperate at the same time?"[1] While asking children what they would like to wear to school can result in nightgowns, costumes, or any of a number of other socially unacceptable possibilities, offering them two options (e.g., the blue or the white dress) allows the child "to feel like she's getting her way, while still keeping you in charge."[2] As one person on the EverydayFamily site commented, "I think that offering choices is a great way to make the toddler feel like they are in control without them actually being in control."[3]

Throughout this book, I have illustrated how various industries now use the "structured choice" as a form of control not just for toddlers but for everyone. I focused on how digital recommendation systems automate this process by guiding users toward certain choices over others. Through this automation, they facilitate larger cultural trends that favor the status quo. Discursively, the notion of choice has gradually become less associated with possibility and liberty and more with anxiety; or, more accurately, choice has become a tool that rearticulates liberty as a burden. These systems treat people like toddlers who may be afraid, unsure, and anxious about the decisions they make and their potential outcomes. Recommendations have thus taken the place of free choice as both a rhetorical and technological tool that guides users' decisions in order to shape their desires and sense of self. Recommendation systems guide people toward normative and standard choices, which are framed as both the safest and the most advantageous for the production of individual happiness rather than liberty.

The companies that rely on digital recommendation systems operate as part of a global culture industry of choice. Theodor Adorno and Max Horkheimer saw the culture industries as standardizing culture through the reproduction of consumable goods that all perpetuate the same normative ideology in support of the domination of the affluent. Rather than simply standardizing processes of production, however, culture industries of choice standardize methods of distribution and, under the guise of enhancing personal decision-making, lead individuals toward conformity and oppression through apparent acts of autonomy. These industries operate through a profound display of doublespeak wherein conformity becomes autonomy, oppression becomes liberty, consumerism becomes empowerment, autonomy becomes collaboration, stagnation becomes happiness, and recommendations become transcendent potentialities.

At the same time, I have emphasized how we far too often mistake recommendations for orders and algorithms as being beyond our control. Like any other form of expression, algorithms and the recommendations that stem from them can be understood, used, and interacted with in a wide variety of ways. Far from simply being accepted at face value, they can be misunderstood, negotiated with, or rejected. With a wide variety of examples, I have illustrated how common (and hilarious) our efforts to challenge and even subvert the wisdom of recommendations (at times called "the wisdom of the crowd") have always been. When we only notice algorithms when they seem to get us right and when we accept their recommendations, we are forgetting the far more numerous times they get us wrong. To appropriate (and butcher) Leo Bersani's famous line, there is a big secret about recommendation technologies: most of them don't work. If you take only one message from this book, I hope it is that algorithms are not omnipotent and we do not have to follow their advice or trust in what they tell us about ourselves or the world around us. This book is, for me, the first step in the long project of both developing methods for critiquing the place of algorithms within our culture and devising tactics and strategies for undoing them.

While recommendation systems can be negotiated with and undermined, I am much less certain as to whether this technology itself can have a net positive benefit, especially in its current forms. Its very existence implies a high degree of advertising, surveillance, energy expenditure, and commodification that all appears antithetical to the growth of a more critical, equitable, and sustainable digital culture and environment. Could these systems be refigured to focus on system needs rather than personal desires? Could they help us find the sublime rather than the familiar? Rather than simply a participatory culture, could they also help develop a participatory democracy? In my most optimistic moments, I hope that by critiquing these systems, we can begin imagining not just how to undo them but also how to repair them in ways that might make the world better.

In my research, I was struck by how many academics have popularized and profited on recommendation systems and how they are transforming academia itself. Indeed, many businesses that rely on recommendation systems, including eHarmony and ModiFace, were started as class or thesis projects by students and their professors. Along with Pattie Maes, the professors at the University of Minnesota's GroupLens Lab continue to use the lab as an incubator for recommendation system applications and companies. Perhaps most famously, both Facebook and Google began as projects by students interested in shaping and managing their own academic cultures.

While academics created recommendation systems to help manage gendered, classed, and racialized expectations and stresses within the contemporary consumer sphere, these systems have also played a crucial role in transforming the academy for neoliberal times. Advertisements for humanities academic positions now routinely ask for scholars who teach, artistically produce, and research a variety of topics in multiple geographic areas and time periods, using both qualitative and quantitative research methods. At this moment, it is hard for me to imagine attempting even a sliver of this work without the constant aid of Google and Amazon's book recommendations. At the same time, universities have heavily supported the growth of the digital humanities, a sector that, while helpfully introducing quantitative analysis and large data sets to the study and pedagogy of literature, arts, and history, has also siphoned off much of the funding of qualitative and critical scholarship in those same areas.[4]

As governments slash the public funding of higher education and administrators, in turn, cut humanities programs, recommendation systems have become a tool that promotes multidisciplinarity as a way to do more with less. I relied heavily on the recommendation systems I studied to guide my research. In particular, Amazon's, Google's, and Google Scholar's search and recommendation algorithms pointed me toward scholarly and popular works on recommendation systems and discussions that I would not have known existed otherwise. With the help of these systems, I pulled information from a diverse set of fields, including psychology, information sciences, sociology, media studies, business, political science, communication studies, film studies, gender studies, computer science, economics, and philosophy, among others. These fields all focus on the relationship between choice, agency, and desire but in very different ways. Without the help of Amazon's recommendations, I would never have made many of the critical links in this work, especially concerning not just the relationship between choice and digital culture but also how that relationship has structured certain forms of feminism and postfeminism. Without Google, I would not have known about the use of recommendation technologies on many websites, including cosmetic surgery sites.

Yet, while I am extremely grateful for these technologies and their help in making this work possible, I also wonder how they imprinted themselves on my approach by making certain evidence easier to locate and thereby making certain critiques more visible than others. Which scholarship, perspectives, and even fields of study did I miss out on because I did not know the correct search terms and because they did not come to me as recommendations? For instance, trying to find art that employs and critiques recommendation systems has been particularly challenging, and I have yet to find the right search term that will guide me toward them. I have struck out with *algorithmic art*, *data art*, *generative art*, *digital art*, *computational art*, and many other terms. Indeed, only by typing in variations on "art made with recommendation systems" was I serendipitously able to come across cosmetic surgeons using these tools—a phenomenon I otherwise may not have connected to these other culture industries of choice or even come across at all. While Google states that its mission "is to organize the world's information and make it universally accessible and useful," this organizing is obviously not ideologically neutral and promotes certain types of access and use over others.[5] With a critical eye, I have relied on these practices of organizing to build my arguments and gain insight into both the problems and the potential of digital recommendation systems.

Throughout this book, I have modeled a critical and qualitative approach to the study of digital technologies and their relationship to contemporary culture. It is my hope that this work will inspire others to interrogate digital technologies of all kinds and encourage them to use and transform them in quirky, wacky, fun, queer, and transgressive ways. While my study has focused primarily on contemporary recommendation systems within a North American context, these technologies are global and my analysis may prove useful for exploring the relationship between choice and technologies in other cultures. At the moment, these systems function primarily within sites of consumerism, but I hope that as artists begin to incorporate them into their work, a more knowing and critical form of recommendation may evolve that allows us to better interrogate whether and how these technologies can be used to fight the often harmful and divisive ideologies that generated them.

Acknowledgments

I would like first to acknowledge that I began writing this book in Los Angeles, California, the traditional lands of the Kizh Kitc Gabrieleños/Tongva. I finished writing it in Edmonton, Alberta, a Treaty 6 territory and traditional gathering place of the Cree, Blackfoot, Métis, Nakota Sioux, Iroquois, Dene, Ojibway/Saulteaux/Anishinaabe, Inuit, and many others whose histories, languages, and cultures continue to influence our vibrant world. Over that time, I have learned how to better respect, enrich, and act as part of the diverse community around me. This is due entirely to the influence of the wonderful people in my life, without whom this work would not exist.

Many people have helped me work through my ideas, suggest research avenues, and challenge my assumptions and analysis throughout this project. Janet Bergstrom, Nick Browne, John Caldwell, Lisa Cartwright, Manohla Dargis, Allyson Field, Dawn Fratini, Jason Gendler, Harrison Gish, Deanna Gross, Devin Hahn, Kelly Haigh, Erin Hill, Maja Manolovic, Jennifer Moorman, Diane Negra, Chon Noriega, David O'Grady, Jennifer Porst, Maya Smukler, Vivian Sobchack, Tess Takahashi, and Janet Walker have all played formative roles in shaping both my scholarship and my sense of what it means to be an academic. I would especially like to thank Kathleen McHugh, who has mentored me throughout my studies and career. With an unreasonable amount of generosity and care (and an entirely reasonable amount of exasperation), she taught me how to write and how to think.

I would also like to thank my many wonderful colleagues, including William Beard, Beth Capper, Russell Cobb, Liz Czach, Elena Del Rio, Astrid Ensslin, Deborah Erkes, Corrinne Harol, Jaymie Heilman, Nat Hurley, Eddy Kent, Sarah Krotz, Michael Litwack, Keavy Martin, Harvey Quamen, Julie Rak, Sharon Romeo, Carolyn Sale, Mark Simpson, Dennis Sweeney, Serra Tinic, and Teresa Zackodnik. Their continual professional and emotional support makes this job possible. I would also like to thank Sue Hamilton, Michael O'Driscoll, and Peter Sinnema for securing funding that helped me complete my research.

In revising this book, I have received a great deal of support from my fabulous editor, Lisa Banning, and everyone at Rutgers University Press. I would also like to thank Chuck Tryon and various anonymous readers who gave me incredible advice at every step of this project.

And finally, I would like to express my endless gratitude to my family for their support and patience. My parents, Janet and Simon Cohn, have always believed in me. My grandparents, Adrian and Barbara Cohn and Angelina and Jim Mollohan, showed me how to be inquisitive, artistic, and kind. My aunt, Sara Cohn, showed me how to be adventurous and independent. My sister Rebecca Cohn helped me with my research by explaining which programs the cool kids were using. With her unending love, generosity, and wisdom, Jaimie Baron is the source of my confidence and the object of my admiration. I have been blessed by the life we share with Kai and Audrey.

Notes

Introduction

Epigraph: John Lee, "Computer to Help Yule Shoppers," *New York Times*, October 25, 1961.

1 Before the 1980s, Neiman Marcus included a hyphen in its name. I am using the current spelling for simplicity.

2 Wanamaker's department store in Philadelphia also reportedly featured a gift advisory system, and it was used in a similar way.

3 Lee, "Computer to Help Yule Shoppers."

4 "In Marketing," *Businessweek*, October 14, 1961, 119.

5 Nancy Hagen, "Santa Claus Has a New Helper," *St. Petersburg (Fla.) Times*, December 24, 1961, Parade.

6 Hagen.

7 Hagen.

8 Lee, "Computer to Help Yule Shoppers."

9 Guy Fedorkow, "About the Computer History Museum's IBM 1401 Machines," Computer History Museum, February 19, 2015, http://www.computerhistory.org /atchm/about-the-computer-history-museums-ibm-1401-machines/.

10 Hagen, "Santa Claus Has a New Helper."

11 Jon Schwarz, "Drones, IBM, and the Big Data of Death," Intercept, October 23, 2015, https://theintercept.com/2015/10/23/drones-ibm-and-the-big-data-of-death/.

12 Paul Waldman, "The Trump Campaign Says It's Suppressing Democratic Votes. Meanwhile Trump Suppresses His Own," *Week*, October 28, 2016, http://theweek .com/articles/658004/trump-campaign-says-suppressing-democratic-votes -meanwhile-trump-suppresses.

13 "Racism Is Poisoning Online Ad Delivery, Says Harvard Professor," *MIT Technology Review*, February 4, 2013, https://www.technologyreview.com/s /510646/racism-is-poisoning-online-ad-delivery-says-harvard-professor/.

14 Samuel Gibbs, "Women Less Likely to Be Shown Ads for High-Paid Jobs on Google, Study Shows," *Guardian*, July 8, 2015, https://www.theguardian.com /technology/2015/jul/08/women-less-likely-ads-high-paid-jobs-google-study.

15 Guillaume Chaslot, "YouTube's A.I. Was Not Neutral in the US Presidential Election," *Medium* (blog), November 27, 2016, https://medium.com/the-graph /youtubes-ai-is-neutral-towards-clicks-but-is-biased-towards-people-and-ideas -3a2f643dea9a.

16 Fruzsina Eordogh, "Why YouTube's Unfixed Restricted Mode Algorithm Is Still the Bigger Story," *Forbes*, April 5, 2017, https://www.forbes.com/sites /fruzsinaeordogh/2017/04/05/why-youtubes-unfixed-restricted-mode-algorithm -is-the-bigger-story/.

17 Rosa Marchitelli, "'Profiting from Hate': Amazon under Fire for Allowing Sale of Nazi Paraphernalia," CBC News, January 8, 2018, http://www.cbc.ca/news /business/amazon-accused-of-profiting-from-hate-1.3358259.

18 Gilles Deleuze, "Society of Control," Nadir, November 25, 1998, http://www .nadir.org/nadir/archiv/netzkritik/societyofcontrol.html.

19 Seb Franklin, *Control: Digitality as Cultural Logic* (Cambridge, Mass.: MIT Press, 2015), xxii.

20 Mark Poster, *The Second Media Age* (New York: John Wiley and Sons, 2013), 91.

21 Lev Manovich, *The Language of New Media* (repr., Cambridge, Mass.: MIT Press, 2002), 61.

22 Louis Althusser quoted in Renata Salecl, *Choice* (London: Profile Books, 2010), 9.

23 Elle Hunt, "The Babadook: How the Horror Movie Monster Became a Gay Icon," *Guardian*, June 11, 2017.

24 For works that frame automated recommendations as a democratizing innovation, see Andrew McAfee and Erik Brynjolfsson, *Machine, Platform, Crowd: Harnessing Our Digital Future* (New York: W. W. Norton, 2017); Hector J. Levesque, *Common Sense, the Turing Test, and the Quest for Real AI* (Cambridge, Mass.: MIT Press, 2017); Arlindo Oliveira, *The Digital Mind: How Science Is Redefining Humanity* (Cambridge, Mass.: MIT Press, 2017); Clay Shirky, *Cognitive Surplus: Creativity and Generosity in a Connected Age* (New York: Penguin, 2010); and Brian Christian and Tom Griffiths, *Algorithms to Live By: The Computer Science of Human Decisions* (repr., New York: Picador, 2017). For works that suggest such recommendations are fascistic and oppressive, see Ed Finn, *What Algorithms Want: Imagination in the Age of Computing* (Cambridge, Mass.: MIT Press, 2017); John Cheney-Lippold, *We Are Data: Algorithms and the Making of Our Digital Selves* (New York: New York University Press, 2017); Safiya Umoja Noble, *Algorithms of Oppression: How Search Engines Reinforce Racism* (New York: New York University Press, 2018); Mark Andrejevic, *Infoglut: How Too Much Information Is Changing the Way We Think and Know* (New York: Routledge, 2013); Cathy O'Neil, *Weapons of Math Destruction: How Big Data Increases Inequality and Threatens Democracy* (repr., New York: Broadway Books, 2017); Frank Pasquale, *The Black Box Society: The Secret Algorithms That Control Money and Information* (Cambridge, Mass.: Harvard University Press, 2015); and Virginia Eubanks, *Automating Inequality: How High-Tech Tools Profile, Police, and Punish the Poor* (New York: St. Martin's, 2018).

25 Doug Turnbull, "Why I Think Search Engines Are the Future of Recommendation Systems," OpenSource Connections, September 13, 2016, http:// opensourceconnections.com/blog/2016/09/13/search-engines-are-the-future-of -recsys/.

26 Matthew Rosenberg, Nicholas Confessore, and Carole Cadwalladr, "How Trump Consultants Exploited the Facebook Data of Millions," *New York Times*,

March 17, 2018, https://www.nytimes.com/2018/03/17/us/politics/cambridge
-analytica-trump-campaign.html.

27 Lisa Gitelman, *Always Already New: Media, History, and the Data of Culture*
(Cambridge, Mass.: MIT Press, 2008), 1.

28 Lisa Gitelman and Geoffrey B. Pingree, introduction to *New Media, 1740–1915*, ed.
Lisa Gitelman and Geoffrey B. Pingree (Cambridge, Mass.: MIT Press, 2003), vi.

29 Pasquale, 1–19.

30 Janet H. Murray, "Inventing the Medium," in *The New Media Reader*, ed. Noah
Wardrip-Fruin and Nick Montfort (Cambridge, Mass.: MIT Press, 2003), 20.

31 Tarleton Gillespie, "Algorithm," in *Digital Keywords: A Vocabulary of Information
Society and Culture*, ed. Benjamin Peters (Princeton, N.J.: Princeton University
Press, 2016), 18.

32 Gillespie, 22.

33 Michael Hickins, "How the NSA Could Get So Smart So Fast," *Wall Street
Journal*, June 12, 2013, https://www.wsj.com/articles/SB10001424127887324049
504578541271020665666; Andrew Leonard, "Netflix, Facebook—and the NSA:
They're All in It Together," Salon, June 14, 2013, https://www.salon.com/2013/06/14
/netflix_facebook_and_the_nsa_theyre_all_in_it_together/.

34 Maggie Zhang, "Google Photos Tags Two African-Americans as Gorillas through
Facial Recognition Software," *Forbes*, July 1, 2015, https://www.forbes.com/sites
/mzhang/2015/07/01/google-photos-tags-two-african-americans-as-gorillas
-through-facial-recognition-software/; Angelique Carson, "Surveillance as a Tool
for Racism," *TechCrunch* (blog), April 25, 2016, http://social.techcrunch.com
/2016/04/25/surveillance-as-a-tool-for-racism/.

35 Mandalit Del Barco, "How Kodak's Shirley Cards Set Photography's Skin-Tone
Standard," NPR, November 13, 2014, https://www.npr.org/2014/11/13/363517842
/for-decades-kodak-s-shirley-cards-set-photography-s-skin-tone-standard.

36 Noble, *Algorithms of Oppression*.

37 James Vincent, "Google 'Fixed' Its Racist Algorithm by Removing Gorillas from
Its Image-Labeling Tech," Verge, January 12, 2018, https://www.theverge.com
/2018/1/12/16882408/google-racist-gorillas-photo-recognition-algorithm-ai.

38 Hilary Brueck, "Is Amazon Prime Too White?," *Fortune*, April 22, 2016,
http://fortune.com/2016/04/22/amazon-prime-in-more-white-neighborhoods/.

39 Gillespie, "Algorithm," 24.

40 Alexander R. Galloway, *Gaming: Essays on Algorithmic Culture*, Electronic
Mediations (Minneapolis: University of Minnesota Press, 2006), 92.

41 Matthew Fuller and Andrew Goffey, *Evil Media* (Cambridge, Mass.: MIT Press,
2012), 76.

42 Scott Kushner, "The Freelance Translation Machine: Algorithmic Culture and
the Invisible Industry," *New Media and Society* 15, no. 8 (2013): 1243.

43 Robert Seyfert and Jonathan Roberge, "What are Algorithmic Cultures?," in
Algorithmic Cultures: Essays on Meaning, Performance and New Technologies, ed.
Robert Seyfert and Jonathan Roberge (New York: Routledge, 2016), 5.

44 Imani Perry, *More Beautiful and More Terrible: The Embrace and Transcendence of
Racial Inequality in the United States* (New York: New York University Press,
2011), 13–23.

45 Tarleton Gillespie, "The Relevance of Algorithms," in *Media Technologies: Essays
on Communication, Materiality, and Society*, ed. Tarleton Gillespie, Pablo J.
Boczkowski, and Kirsten A. Foot (Cambridge, Mass.: MIT Press, 2014), 168.

46 Siva Vaidhyanathan, *The Googlization of Everything*, updated ed. (Berkeley: University of California Press, 2012), 1.

47 Finn, *What Algorithms Want*, 8.

48 Dietmar Jannach et al., *Recommender Systems: An Introduction* (New York: Cambridge University Press, 2011), 1.

49 Jannach, et al., 2.

50 David Goldberg et al., "Using Collaborative Filtering to Weave an Information Tapestry," *Communications of the ACM* 35, no. 12 (1992): 61–70.

51 Goldberg et al., 60.

52 Goldberg et al., 61.

53 Peter Kafka, "Did Amazon Really Fail This Weekend? The Twittersphere Says 'Yes,' Online Retailer Says 'Glitch,'" AllThingsD, April 12, 2009, http://allthingsd.com/20090412/did-amazon-really-fail-this-weekend-the-twittersphere-says-yes/.

54 Peter Kafka, "Amazon Apologizes for 'Ham-Fisted Cataloging Error,'" AllThingsD, April 13, 2009, http://allthingsd.com/20090413/amazon-apologizes-for-ham-fisted-cataloging-error/.

55 M. Carrier, "Strange Amazon Recommendations," Amazon, accessed December 1, 2012, https://web.archive.org/web/20140418013307/http://www.amazon.com/Strange-Amazon-Recommendations/lm/R15W3LLZV2GF0V.

56 Charles Duhigg, "How Companies Learn Your Secrets," *New York Times*, February 16, 2012, http://www.nytimes.com/2012/02/19/magazine/shopping-habits.html.

57 Christian Sandvig, "Corrupt Personalization," Social Media Collective (blog), June 26, 2014, https://socialmediacollective.org/2014/06/26/corrupt-personalization/

58 Steve Lohr, "Big Data Is Great, but Don't Forget Intuition," *New York Times*, December 29, 2012, http://www.nytimes.com/2012/12/30/technology/big-data-is-great-but-dont-forget-intuition.html.

59 J. P. Mangalindan, "Amazon's Recommendation Secret," *Fortune*, July 30, 2012, http://fortune.com/2012/07/30/amazons-recommendation-secret/; Shabana Arora, "Recommendation Engines: How Amazon and Netflix Are Winning the Personalization Battle," Martech Advisor, June 28, 2016, https://www.martechadvisor.com/articles/customer-experience/recommendation-engines-how-amazon-and-netflix-are-winning-the-personalization-battle/; Tom Krawiec, "Amazon's Recommendation Engine: The Secret to Selling More Online," Rejoiner, June 6, 2016, http://rejoiner.com/resources/amazon-recommendations-secret-selling-online/.

60 Nancy Buckley, "Neiman Marcus Launches MyNM to Personalize Ecommerce," Luxury Daily, October 27, 2014, https://www.luxurydaily.com/neiman-marcus-launches-mynm-to-personalize-online-shopping/.

61 Phil Wahba, "Barneys E-Commerce Sites Could Give Neiman, Nordstrom Agita," *Fortune*, March 3, 2015, http://fortune.com/2015/03/03/barneys-new-york-e-commerce/.

62 Buckley, "Neiman Marcus Launches MyNM."

Chapter 1   A Brief History of Good Choices

1 Jean-Paul Sartre, *Existentialism Is a Humanism*, trans. Carol Macomber (New Haven, Conn.: Yale University Press, 2007), 33.

2 Stephen Priest, "Sartre in the World," in *Jean-Paul Sartre: Basic Writings*, ed. Stephen Priest (London: Routledge, 2000), 16.

3 Sartre, *Existentialism Is a Humanism*, 10.

4 Linda Duits and Liesbet van Zoonen, "Headscarves and Porno-Chic: Disciplining Girls' Bodies in the European Multicultural Society," *European Journal of Women's Studies* 13, no. 2 (2006): 103–117.

5 Raymond Williams, *Keywords: A Vocabulary of Culture and Society*, rev. ed. (New York: Oxford University Press, 1985), 22.

6 Laura Stampler, "A Professional Binge Watcher Explains What It's Like to Get Paid to Watch Netflix," *Time*, July 14, 2014, http://time.com/2970271/job-netflix -professional-binge-watcher/.

7 *Oxford English Dictionary Online*, s.v. "suggestion (*n.*)," accessed March 11, 2017, http://www.oed.com/view/Entry/193668.

8 John Milton, *Samson Agonistes: A Dramatic Poem* (CreateSpace Independent Publishing Platform, 2017), lines 599–600.

9 Geoffrey Chaucer, *The Parson's Tale* (Arcadia, Calif.: Tumblar House, 2011), line 330; Thomas Kyd, *The Spanish Tragedy* (CreateSpace Independent Publishing Platform, 2011), 1.3.1.

10 *Oxford English Dictionary Online*, s.v. "suggestion (*n.*)."

11 Sheridan quoted in Robert Booth, "Facebook Reveals News Feed Experiment to Control Emotions," *Guardian*, June 30, 2014, https://www.theguardian.com /technology/2014/jun/29/facebook-users-emotions-news-feeds.

12 Booth.

13 Ann Coulter, "How the Media Work—Breitbart," Breitbart, August 17, 2016, http://www.breitbart.com/big-journalism/2016/08/17/ann-coulter-media-work/.

14 *Holy Bible Dictionary/Concordance*, Red Letter ed., King James Version (World, 1986), Acts 14:26, 15:40.

15 *Oxford English Dictionary Online*, "recommendation (*n.*)," accessed March 11, 2017, http://www.oed.com/view/Entry/159718.

16 Jane Austen, *Pride and Prejudice*, Penguin Classics ed. (London Penguin Books, 2002), 93.

17 Editorial staff, "Gwyneth Paltrow Just Endorsed a $15,000 Golden Dildo," *NextShark* (blog), May 10, 2016, https://nextshark.com/gwyneth-paltrow-dildo -golden-lelo/.

18 Nicholas Negroponte, *Being Digital* (New York: Knopf Doubleday, 2015), 153–154.

19 Joseph Turow, *The Daily You: How the New Advertising Industry Is Defining Your Identity and Your Worth* (New Haven, Conn.: Yale University Press, 2013), 8; Cass R. Sunstein, *#Republic: Divided Democracy in the Age of Social Media* (Princeton, N.J.: Princeton University Press, 2018).

20 Kaveh Waddell, "How Algorithms Can Bring Down Minorities' Credit Scores," *Atlantic*, December 2, 2016, https://www.theatlantic.com/technology/archive /2016/12/how-algorithms-can-bring-down-minorities-credit-scores/509333/.

21 Julia Angwin et al., "Machine Bias: There's Software Used across the Country to Predict Future Criminals. And It's Biased against Blacks," ProPublica, May 23, 2016, https://www.propublica.org/article/machine-bias-risk-assessments-in -criminal-sentencing.

22 Sapna Maheshwari, "On YouTube Kids, Startling Videos Slip Past Filters," *New York Times*, November 4, 2017, https://www.nytimes.com/2017/11/04/business /media/youtube-kids-paw-patrol.html.

23 "Welcome to DashVapes," DashVapes, accessed March 11, 2017, https://www
.dashvapes.com/guru.
24 Judith Butler, "Sex and Gender in Simone de Beauvoir's Second Sex," *Yale French
Studies* 72 (Winter 1986): 37.
25 Butler, 36.
26 Butler, 40.
27 Ed Finn, *What Algorithms Want: Imagination in the Age of Computing* (Cambridge, Mass.: MIT Press, 2017), 31–34;Tama Leaver, *Artificial Culture: Identity,
Technology, and Bodies* (New York: Routledge, 2014), 7.
28 Alan Turing, "Computing Machinery and Intelligence," *Mind* 59 (1950): 433.
29 N. Katherine Hayles, *How We Became Posthuman: Virtual Bodies in Cybernetics,
Literature, and Informatics* (Chicago: University of Chicago Press, 1999), xi.
30 Turing; Norbert Wiener, *The Human Use of Human Beings: Cybernetics and
Society* (Boston: Da Capo Press, 1988.
31 Angela McRobbie, *The Aftermath of Feminism: Gender, Culture and Social
Change* (Los Angeles: SAGE, 2009); Susan Faludi, *Backlash: The Undeclared War
against American Women* (New York: Crown, 1991).
32 David Hesmondhalgh, *The Cultural Industries*, 3rd ed. (Thousand Oaks, Calif.:
Sage, 2012), 556.
33 Mimi White, *Tele-advising: Therapeutic Discourse in American Television* (Chapel
Hill: University of North Carolina Press, 1992), 7.
34 White, 9.
35 "Eliza Chat Bot," nlp-addiction.com, accessed July 21, 2015, https://web.archive
.org/web/20150314003348/http://nlp-addiction.com/eliza/; Michal Wallace,
"Eliza, Computer Therapist," Manifestation.com, 2006, http://www
.manifestation.com/neurotoys/eliza.php3.
36 Sarah Nilsen and Sarah E. Turner, introduction to *The Colorblind Screen:
Television in Post-racial America*, ed. Sarah Nilsen and Sarah E. Turner (New
York: New York University Press, 2014), 2.
37 Imani Perry, *More Beautiful and More Terrible: The Embrace and Transcendence of
Racial Inequality in the United States* (New York: New York University Press,
2011), 85–126.
38 Catherine R. Squires, *The Post-racial Mystique: Media and Race in the Twenty-
First Century* (New York: New York University Press, 2014), 6.
39 Rosalind C. Gill, "Critical Respect: The Difficulties and Dilemmas of Agency and
'Choice' for Feminism: A Reply to Duits and van Zoonen," *European Journal of
Women's Studies* 14, no. 1 (February 1, 2007): 71.
40 Gill, 72.
41 Stéphanie Genz and Benjamin A. Brabon, *Postfeminism: Cultural Texts and
Theories* (Edinburgh: Edinburgh University Press, 2009), 171.
42 Yvonne Tasker and Diane Negra, "Introduction: Feminist Politics and Postfeminist Culture," in *Interrogating Postfeminism: Gender and the Politics of Popular
Culture*, ed. Yvonne Tasker and Diane Negra (Durham, N.C.: Duke University
Press, 2007), 2.
43 Rachel Bertsche. "I Chose to Be a Stay-At-Home Wife," *Cosmopolitan*, April 8,
2014. http://www.cosmopolitan.com/sex-love/relationship-advice/stay-at-home
-wife-essay.
44 Paul Resnick et al., "GroupLens: An Open Architecture for Collaborative
Filtering of Netnews," *Proceedings of the 1994 ACM Conference on Computer
Supported Cooperative Work* (1994): http://ccs.mit.edu/papers/CCSWP165.html.

45 Resnick et al.
46 Resnick et al.
47 Resnick et al.
48 Eli Pariser, *The Filter Bubble: What the Internet Is Hiding from You* (New York: Penguin, 2011).
49 Joseph Konstan, interview by author, Skype, July 5, 2016.
50 "Berkeley Workshop on Collaborative Filtering Conference Program," March 16, 1996, http://www2.sims.berkeley.edu/resources/collab/.
51 Konstan, interview.
52 "Fine Tuning the Social Web: John Riedl," University of Minnesota College of Science and Engineering blog, April 22, 2013, https://research.umn.edu/inquiry/post/fine-tuning-social-web-john-riedl.
53 John Riedl and Joseph Konstan, *Word of Mouse: The Marketing Power of Collaborative Filtering*, with Eric Vrooman (New York: Business Plus, 2002), 109.
54 Malcolm Gladwell, "The Science of the Sleeper," *New Yorker*, October 4, 1999, http://www.newyorker.com/magazine/1999/10/04/the-science-of-the-sleeper.
55 Wendy Hui Kyong Chun, *Updating to Remain the Same: Habitual New Media* (Cambridge, MA: MIT Press, 2016), 54.
56 Riedl and Konstan, *Word of Mouse*, 110.
57 Riedl and Konstan, 110.
58 Chun, *Updating to Remain the Same*, 40.
59 Joseph Konstan, "Lecture 9: Recommender Systems: Past, Present and Future—University of Minnesota," Coursera, accessed March 11, 2017, https://www.coursera.org/learn/recommender-systems-introduction/lecture/GqB2e/recommender-systems-past-present-and-future.
60 Konstan, interview.
61 Riedl and Konstan, *Word of Mouse*, 1.
62 Riedl and Konstan, 207.
63 Riedl and Konstan, 223.
64 "MoMa Integrates Certona Personalization with a Site Rebuild & Sees It Contribute 12% of Total Site Revenue," Certona, December 2015, http://www.certona.com/wp-content/uploads/2015/12/Certona-Client-Success-MoMA-new.pdf.
65 "Customers," Barilliance, 2013, http://www.barilliance.com/customers.
66 "Products & Solutions," Baynote, 2013, http://www.baynote.com/products-solutions/.
67 Karl Polyani quoted in David Harvey, *A Brief History of Neoliberalism* (New York: Oxford University Press, 2007), 38.
68 Tasker and Negra, "Introduction," 1.
69 Harvey, *Brief History of Neoliberalism*, 65.
70 Renata Salecl, *Choice* (London: Profile Books, 2010), 60.
71 Jean-Paul Sartre, *Being and Nothingness*, trans. Hazel E. Barnes (repr., New York: Washington Square, 1993), 101.
72 Sartre, 102.
73 Sheena Iyengar, *The Art of Choosing* (New York: Twelve, 2010); Sheena Iyengar and Mark Lepper, "When Choice is Demotivating: Can One Desire Too Much of a Good Thing?" *Journal of Personality and Social Psychology*, 79, no. 6 (2001): 995–1006.
Barry Schwartz, *The Paradox of Choice: Why More Is Less* (New York: Ecco, 2004).

74 Salecl, *Choice*, 1.
75 Salecl, 4.
76 Max Horkheimer and Theodor W. Adorno, *Dialectic of Enlightenment: Philosophical Fragments,* ed. Gunzelin Schmid Noerr, trans. Edmund Jephcott (Stanford, Calif.: Stanford University Press, 2002), 97.
77 Riedl and Konstan, *Word of Mouse,* 238.
78 Judith Halberstam, *The Queer Art of Failure* (Durham, N.C.: Duke University Press, 2011), 2.
79 Halberstam, 4.
80 Halberstam, 91.
81 Sarah Ahmed, *The Promise of Happiness* (Durham N.C.: Duke University Press, 2010); Jasbir K. Puar, *Terrorist Assemblages: Homonationalism in Queer Times* (Durham, N.C.: Duke University Press, 2007).
82 Kara Keeling, *The Witch's Flight: The Cinematic, the Black Femme, and the Image of Common Sense* (Durham, N.C.: Duke University Press, 2007), 15.
83 Keeling, 4.
84 Sara Ahmed, *Queer Phenomenology: Orientations, Objects, Others* (Durham, N.C.: Duke University Press, 2006), 189; Jasbir K. Puar, *Terrorist Assemblages: Homonationalism in Queer Times* (Durham, N.C.: Duke University Press, 2007), 203–223.

Chapter 2   Female Labor and Digital Media

1 See N. Katherine Hayles, *My Mother Was a Computer: Digital Subjects and Literary Texts* (Chicago: University of Chicago Press, 2005); Wendy Hui Kyong Chun, *Programmed Visions: Software and Memory,* Software Studies (Cambridge, Mass.: MIT Press, 2011); and Sadie Plant, *Zeroes and Ones: Digital Women and the New Technoculture* (New York: Doubleday, 1997).
2 Anne Marie Balsamo, *Designing Culture: The Technological Imagination at Work* (Durham, N.C.: Duke University Press, 2011), 32.
3 During this period, work on collaborative filtering and recommendation systems continued at several other corporations and university departments. Along with GroupLens, which I discuss in chapter 1, labs at Bell Communications Research in New Jersey, Xerox's Palo Alto Research Center in California, and the International Computer Science Institute at the University of California, Berkeley, worked on collaborative filtering in parallel to Maes. Major media outlets were also interested in creating recommendation systems to help sell their products. Paramount Interactive created a movie recommendation software package called Movie Select. Like Firefly, this product attempted to recommend films that users might be interested in. The main difference between these technologies was that while Firefly used a data set (list of movies and media) that was constantly expanding and changing, the Paramount lists were static and stored "correlations between different items and [used] those correlations to make recommendations. As such, their recommendations are less personalized than in social information filtering systems." See Upendra Shardanand and Pattie Maes. "Social Information Filtering: Algorithms for Automating 'Word of Mouth,'" *ACM CHI '95 Proceedings,* 1995. http://prior.sigchi.org/chi95/Electronic /documnts/papers/us_bdy.htm. This is not to say that these various other companies did not contribute a great deal to the development of these technolo-

gies; rather, they were never the focus of these media conglomerates' attention, and their work was often considered secondary to that done at MIT. They also did not incorporate social networking into their sites, which proved to be central to Firefly's success.

4 In contrast to her colleagues, Maes never wrote books for a popular audience, a possible explanation for her lack of fame.

5 Yvonne Tasker and Diane Negra, "Introduction: Feminist Politics and Postfeminist Culture," in *Interrogating Postfeminism: Gender and the Politics of Popular Culture*, ed. Yvonne Tasker and Diane Negra (Durham, N.C.: Duke University Press, 2007), 1.

6 Angela McRobbie, *The Aftermath of Feminism: Gender, Culture and Social Change* (Los Angeles: SAGE, 2009); Susan Faludi, *Backlash: The Undeclared War against American Women* (New York: Crown, 1991).

7 See Lisa Duggan, *The Twilight of Equality? Neoliberalism, Cultural Politics, and the Attack on Democracy* (Boston: Beacon, 2003); and McRobbie, *Aftermath of Feminism*.

8 Tracie Egan Morrissey, "'Feminist Housewife' Writer Lisa Miller Is Trying to Make 'Lean Out' Happen," Jezebel, March 21, 2013, https://jezebel.com/5991589 /feminist-housewife-writer-lisa-miller-is-trying-to-make-lean-out-happen.

9 Doug Barry, "Moms Don't 'Opt Out' of Work Because They're Super Wealthy," Jezebel, May 12, 2013, http://jezebel.com/moms-don-t-opt-out-of-work-because -they-re-super-weal-503306759.

10 Diane Negra, *What a Girl Wants? Fantasizing the Reclamation of Self in Postfeminism* (London: Routledge, 2009), 4.

11 Tasker and Negra, "Introduction," 2.

12 Negra, *What a Girl Wants?*, 3–4.

13 Tasker and Negra, "Introduction," 4.

14 As I write this in the midst of both various misogynistic rants from President Donald J. Trump and women's marches and #MeToo and #TimesUp campaigns against sexual harassment, it is clear that feminism not only is alive and well but has perhaps always existed in conjunction with a postfeminist discourse and logic; one does not preclude the other. This conjunction is now demonstrated by continual debates over whether #MeToo would have gotten any traction had it not been for the numerous white celebrities willing to share their experiences. And while Trump, for good reason, is often considered the prime catalyst for this revitalization of feminism, this ironically has had the effect of obscuring the years of activism that made these campaigns not just successful but possible: feminism and postfeminism do not have a causal or linear relationship but are rather intertwined.

15 Tasker and Negra, "Introduction," 10.

16 McRobbie, *Aftermath of Feminism*, 11–24.

17 Steve W. Rawlings, "Population Profile of the United States: Households and Families" (Washington, D.C., 1994), https://web.archive.org/web/2004022302 0730/http://www.census.gov/population/www/pop-profile/hhfam.html.

18 Rosalind Gill and Christina Scharff, introduction to *New Femininities: Postfeminism, Neoliberalism, and Subjectivity*, ed. Rosalind Gill and Christina Scharff (Basingstoke, U.K.: Palgrave Macmillan, 2011), 7.

19 Steve Cohan, "Queer Eye for the Straight Guise: Camp, Postfeminism and the Fab Five's Makeovers of Masculinity," in Tasker and Negra, *Interrogating Postfeminism*, 176–201.

20 Sheena Iyengar, *The Art of Choosing* (New York: Twelve, 2010), 187.
21 Iyengar, 188.
22 Iyengar, 203.
23 Iyengar, 205.
24 This project was featured in the Errol Morris documentary *Fast, Cheap and Out of Control* (1997), which focused on the similarities between contemporary human-ity and these simple robots. Ironically, while Maes used human biology to create the models for AI, Morris uses Maes's and Brooks's AI work as an analogy for human identity and individuality.
25 Mary C. Boyce et al., *Report of the Committee on Women Faculty in the School of Engineering at MIT* (Boston: Massachusetts Institute of Technology, March 2002), 11.
26 Pattie Maes, interview by the author, email, June 21, 2011.
27 Maes, interview.
28 Maes.
29 Maes.
30 Maes.
31 Sallie W. Chisholm et al., *A Study on the Status of Women Faculty in Science at MIT* (Boston: Massachusetts Institute of Technology, 1999).
32 Jeffrey Mervis, "MIT Pledges to Improve Conditions for Minority Faculty Members," ScienceInsider, January 14, 2010, http://news.sciencemag.org/2010/01/mit-pledges-improve-conditions-minority-faculty-members.
33 Sherry Turkle, *Life on the Screen: Identity in the Age of the Internet* (New York: Simon and Schuster, 1995), 99.
34 Nicholas Negroponte and Pattie Maes, "Electronic Word of Mouth," *Wired*, October 1996.
35 Jennifer Baxter, "Stay-at-Home Dads," Australian Institute of Family Studies, May 2017, https://aifs.gov.au/publications/stay-home-dads.
36 Negroponte and Maes, "Electronic Word of Mouth."
37 Turkle, *Life on the Screen*, 99.
38 Negroponte and Maes.
39 Turkle, *Life on the Screen*, 100.
40 Tomoko Koda and Pattie Maes, "Agents with Faces: The Effects of Personification of Agents," *Proceedings of HCI '96*. 1996, 6.
41 Turkle, *Life on the Screen*, 100.
42 Lev Manovich, *The Language of New Media* (repr., Cambridge, Mass.: MIT Press, 2002), 61.
43 Hugo Liu, Pattie Maes, and Glorianna Davenport. "Unraveling the Taste Fabric of Social Networks," *International Journal on Semantic Web and Information Systems* 2, no. 1 (2006): 42–71.
44 Liu, Maes, and Davenport, 46.
45 See McRobbie, *Aftermath of Feminism*.
46 Mark Andrejevic, *Infoglut: How Too Much Information Is Changing the Way We Think and Know* (New York: Routledge, 2013), 8.
47 Anna Everett, "Digitextuality and Click Theory: Theses on Convergence Media in the Digital Age," in *New Media: Theories and Practices of Digitextuality*, ed. John Thornton Caldwell and Anna Everett, AFI Film Readers (New York: Routledge, 2003), 14.
48 Liu, Maes, and Davenport, 44.

49 For a small selection of the range of theorists who argue this, see Charles Taylor, *Sources of the Self: The Making of the Modern Identity* (Cambridge, Mass.: Harvard University Press, 1989); and Scott Bukatman, *Terminal Identity: The Virtual Subject in Postmodern Science Fiction* (Durham, N.C.: Duke University Press, 1993).

50 Liu, Maes, and Davenport, 46.

51 Liu, Maes, and Davenport, 47.

52 Liu, Maes, and Davenport, 45.

53 Liu, Maes, and Davenport, 46.

54 Liu, Maes, and Davenport, 46.

55 Timothy O'Connor and Constantine Sandis, *A Companion to the Philosophy of Action*, Blackwell Companions to Philosophy (Chichester, U.K.: Wiley-Blackwell, 2010), 280.

56 O'Connor and Sandis, 274.

57 Donna J. Haraway, "A Manifesto for Cyborgs: Science, Technology and Socialist Feminism," in *Liquid Metal: The Science Fiction Film Reader*, ed. Sean Redmond (New York: Wallflower, 2004), 158–181.

58 N. Katherine Hayles, *How We Became Posthuman: Virtual Bodies in Cybernetics, Literature, and Informatics* (Chicago: University of Chicago Press, 1999), 19.

59 Liu, Maes, and Davenport, "Unraveling the Taste Fabric," 45.

60 Liu, Maes, and Davenport, 45.

61 Liu, Maes, and Davenport, 45.

62 Jasbir K. Puar, *Terrorist Assemblages: Homonationalism in Queer Times* (Durham, N.C.: Duke University Press, 2007), 197.

63 Horace Campbell quoted in Puar, 197.

64 Puar, 197.

65 Negroponte and Maes, "Electronic Word of Mouth."

66 Maes, interview.

67 Chuck Tryon, *On-Demand Culture: Digital Delivery and the Future of Movies* (New Brunswick, N.J.: Rutgers University Press, 2013), 115–124.

68 Shardanand and Maes, "Social Information Filtering."

69 Pattie Maes founded Agents with Upendra Shardanand, Nick Grouf, Max Metral, David Waxman, and Yezdi Lashkari.

70 "Prix Ars Electronica 1995 World Wide Web Distinction: Ringo++," Prix Ars Electronica, 1995, http://90.146.8.18/en/archives/prix_archive/prix_projekt.asp?iProjectID=2529.

71 Maes, interview.

72 Maes, interview.

73 For a sample of these discussions, see many of the articles located in Tasker and Negra, *Interrogating Postfeminism*.

74 Maes, interview. See also Leonard Newton Foner, "Political Artifacts and Personal Privacy: The Yenta Multi-agent Distributed Matchmaking System" (PhD diss., Massachusetts Institute of Technology, 1999), https://dspace.mit.edu/handle/1721.1/61104.

75 Pattie Maes, "Intelligent Software," *Scientific American*, 150th anniversary issue, September 1995.

76 Maes.

77 Maes.

78 David Harvey, *A Brief History of Neoliberalism* (New York: Oxford University Press, 2007), 116.
79 Frank Pasquale, *The Black Box Society: The Secret Algorithms That Control Money and Information* (Cambridge, Mass.: Harvard University Press, 2015).
80 Turkle, *Life on the Screen*, 100.
81 Pat Hensley et al., *Implementation of OPS over HTTP* (Washington, D.C.: World Wide Web Consortium, June 2, 1997), https://www.w3.org/TR/NOTE-OPS-OverHTTP.
82 Hensley et al.
83 Hensley et al.
84 Hensley et al.
85 Hensley et al.
86 Pattie Maes, "Curriculum Vitae of Pattie Maes," 1997, https://web.archive.org/web/20170728042632/http://web.media.mit.edu:80/~pattie/cv.html.
87 "Pattie Maes," *People*, May 12, 1997. In response to the article, Maes emailed her department to explain the misquote and then posted an explanation on her homepage.
88 Negra, *What a Girl Wants?*, 4.

Chapter 3   Mapping the Stars

1 "Hacking Creativity," Red Bull High Performance Group, accessed December 1, 2017, http://hackingcreativity.com/.
2 David Carr, "For 'House of Cards,' Using Big Data to Guarantee Its Popularity," *New York Times*, February 24, 2013, http://www.nytimes.com/2013/02/25/business/media/for-house-of-cards-using-big-data-to-guarantee-its-popularity.html.
3 Neil Andreeva, "Netflix to Enter Original Programming with Mega Deal for David Fincher-Kevin Spacey Series 'House of Cards,'" Deadline, March 15, 2011, http://deadline.com/2011/03/netflix-to-enter-original-programming-with-mega-deal-for-david-fincher-kevin-spacey-drama-series-house-of-cards-114184/.
4 Chuck Tryon, "TV Got Better: Netflix's Original Programming Strategies and Binge Viewing," *Media Industries Journal* 2, no. 2 (2015), http://dx.doi.org/10.3998/mij.15031809.0002.206.
5 Susan Ohmer, *George Gallup in Hollywood* (New York: Columbia University Press, 2006).
6 Timothy Havens, "Media Programming in an Era of Big Data," *Media Industries Journal* 1, no. 2 (2014), http://dx.doi.org/10.3998/mij.15031809.0001.202.
7 Frank Pasquale, *The Black Box Society: The Secret Algorithms That Control Money and Information* (Cambridge, Mass.: Harvard University Press, 2015).
8 John Cheney-Lippold, *We Are Data: Algorithms and the Making of Our Digital Selves* (New York: New York University Press, 2017), 1–37.
9 Ed Finn, *What Algorithms Want: Imagination in the Age of Computing* (Cambridge, Mass.: MIT Press, 2017), 2.
10 Jacques Derrida, *Of Grammatology*, trans. Gayatri Chakravorty Spivak (Baltimore: Johns Hopkins University Press, 1998), 3.
11 Gayatri Chakravorty Spivak, introduction to Derrida, *Of Grammatology*, lxviii.
12 Derrida, *Of Grammatology*, 167.
13 Jacques Derrida, *Limited Inc* (Evanston, Ill.: Northwestern University Press, 1977), 236.

14 N. Katherine Hayles, *How We Became Posthuman: Virtual Bodies in Cybernetics, Literature, and Informatics* (Chicago: University of Chicago Press, 1999), 285.

15 Hayles, 285.

16 Hayles, 285.

17 Cheney-Lippold, *We Are Data*, 128–167.

18 Cheney-Lippold, 161.

19 Jack Reynolds, "Jacques Derrida (1930–2004)," Internet Encyclopedia of Philosophy, accessed December 4, 2017, http://www.iep.utm.edu/derrida /#SH2a.

20 David Weinberger, "Our Machines Now Have Knowledge We'll Never Understand," *Wired*, April 18, 2017, https://www.wired.com/story/our-machines-now -have-knowledge-well-never-understand/.

21 Maurice Merleau-Ponty, "The Algorithm and the Mystery of Language," in *Maurice Merleau-Ponty: Basic Writings* (New York: Psychology Press, 2004), 234–246; Michel Serres, *The Five Senses: A Philosophy of Mingled Bodies* (London: Bloomsbury, 2016), 344.

22 Finn, *What Algorithms Want*, 8.

23 Michael Dolan, "The Bork Tape Saga," 2005, https://web.archive.org/web /20160312032732/http://theamericanporch.com/bork4.htm.

24 Dolan.

25 Dolan.

26 Tania Modleski, *The Women Who Knew Too Much: Hitchcock and Feminist Theory* (New York: Routledge, 2015).

27 Dolan, "Bork Tape Saga."

28 Chuck Tryon, *On-Demand Culture: Digital Delivery and the Future of Movies* (New Brunswick, N.J.: Rutgers University Press, 2013), 116.

29 See Glee Harrah Cady and Pat McGregor, "Protect Your Digital Privacy: Survival Skills for the Information Age," in *Protect Your Digital Privacy! Survival Skills for the Information Age* (Indianapolis: Que, 2002), 375; "Video Privacy Protection Act," Electronic Privacy Information Center, accessed August 29, 2018, http:// epic.org/privacy/vppa/; and Joe Biden, *Committee on the Judiciary Report on the Video Privacy Protection Act,* Report 100-599 (Washington D.C.: Government Printing Office, 1988), https://epic.org/privacy/vppa/Senate-Report-100-599.pdf.

30 Biden, 88.

31 Biden, 89.

32 Biden, 115.

33 Biden, 115.

34 Biden, 115.

35 Biden, 115.

36 Biden, 117.

37 "Video Privacy Protection Act."

38 Ryan Singel, "Netflix Spilled Your Brokeback Mountain Secret, Lawsuit Claims," *Wired*, December 17, 2009.

39 Finn, *What Algorithms Want*, 9.

40 Much of this section is revised and updated from Jonathan Cohn, "My TiVo Thinks I'm Gay: Algorithmic Culture and Its Discontents," *Television and New Media* 17, no. 8: 1–16. Copyright © Jonathan Cohn. Reprinted by permission of SAGE Publications, http://journals.sagepub.com/doi/abs/10.1177 /1527476416644978.

41 Cheney-Lippold, *We Are Data*, 33.
42 "Weird Al" Yankovic, "Couch Potato," track 1 on *Poodle Hat*, Volcano Entertainment, 2003.
43 Lisa Parks, "Flexible Microcasting: Gender, Generation, and Television-Internet Convergence," in *Television after TV: Essays on a Medium in Transition*, ed. Lynn Spigel and Jan Olsson (Durham, N.C.: Duke University Press, 2004), 135.
44 Parks, 135.
45 William Boddy, "Interactive Television and Advertising Form in Contemporary U.S. Television," in Spigel and Olsson, *Television after TV*, 113–132.
46 "Average Number of TV Channels Receivable by US Household Drops in 2004," ZDNet, September 29, 2004, http://www.zdnet.com/article/average-number-of-tv-channels-receivable-by-us-household-drops-in-2004/.
47 "Tivo Network Executives Commercial," Goodby Silverstein and Partners, 2000, video, 0:30, https://www.youtube.com/watch?v=f2X1p0sX9CM.
48 Max Robins, "To TiVo or Not to TiVo," *Broadcasting and Cable*, October 3, 2005.
49 Ron Becker, *Gay TV and Straight America* (New Brunswick, N.J.: Rutgers University Press, 2006), 5.
50 Becker, 81.
51 Becker, 104.
52 Becker, 119.
53 Becker, 128.
54 Lisa Henderson, *Love and Money: Queers, Class, and Cultural Production* (New York: New York University Press, 2013), 34.
55 Henderson, 41.
56 Becker, *Gay TV*, 185.
57 Becker, 4.
58 *The Mind of the Married Man*, season 2, episode 2, "The Cream of the Crop," aired Sept. 22, 2002, on HBO.
59 *King of Queens*, season 5, episode 5, "Mammary Lane," aired Oct. 21, 2002, on CBS.
60 Henderson, *Love and Money*, 16.
61 Michael Z. Newman and Elana Levine, *Legitimating Television: Media Convergence and Cultural Status* (New York: Routledge, 2011), 71.
62 Newman and Levine, 1.
63 Peter Yang, "RuPaul: The King of Queens," *Rolling Stone*, October 4, 2013, http://www.rollingstone.com/movies/news/rupaul-the-king-of-queens-20131004.
64 Jeffrey Zaslow, "If TiVo Thinks You Are Gay, Here's How to Set It Straight," *Wall Street Journal*, November 26, 2002.
65 Zaslow.
66 Zaslow.
67 Zaslow.
68 Zaslow.
69 Zaslow.
70 Judiciary Committee, "Video Privacy Protection Act," 131.
71 Kevin McDonald, "Digital Dreams in a Material World: The Rise of Netflix and Its Impact on Changing Distribution and Exhibition Patterns," *Jump Cut* 55 (2013), http://ejumpcut.org/archive/jc55.2013/McDonaldNetflix/index.html.

72 Jim Bennett, "The Cinematch System: Operation, Scale, Coverage, Accuracy Impact" (slideshow, Netflix, September 13, 2006), http://www.slideshare.net /rzykov/netflix-1542324.

73 "Netflix Consumer Press Kit," Netflix, April 25, 2011.

74 Tom Vanderbilt, "The Science behind the Netflix Algorithms That Decide What You'll Watch Next," *Wired*, August 7, 2013, http://www.wired.com/2013/08/qq _netflix-algorithm/.

75 Vanderbilt.

76 Steve Lohr, "Netflix Awards $1 Million Prize and Starts a New Contest," *New York Times*, September 21, 2009.

77 Bennett, "Cinematch System."

78 Bennett.

79 Xavier Amatriain, "Netflix Recommendations: Beyond the 5 Stars (Part 1)," *Medium* (blog), April 6, 2012, https://medium.com/netflix-techblog/netflix -recommendations-beyond-the-5-stars-part-1-55838468f429.

80 Rani Molla, "Netflix Now Has Nearly 118 Million Streaming Subscribers Globally," *Recode*, January 22, 2018, https://www.recode.net/2018/1/22/16920150 /netflix-q4-2017-earnings-subscribers.

81 Bennett.

82 Lohr, "Netflix Awards $1 Million."

83 Lohr.

84 Lohr; Eliot van Buskirk, "How the Netflix Prize Was Won," *Wired*, September 22, 2009, http://www.wired.com/business/2009/09/how-the-netflix-prize -was-won/.

85 John Thornton Caldwell, *Production Culture: Industrial Reflexivity and Critical Practice in Film and Television*, Console-Ing Passions (Durham, N.C.: Duke University Press, 2008); Mark Deuze, *Media Work* (Malden, Mass.: Polity, 2007); Jonathan Cohn, "All Work and No Play," in *Sampling Media*, ed. David Laderman and Laurel Westrup (Oxford: Oxford University Press, 2014), 184–196.

86 Blake Hallinan and Ted Striphas, "Recommended for You: The Netflix Prize and the Production of Algorithmic Culture," *New Media and Society* 18, no. 1 (2016): 117–137.

87 Clive Thompson, "If You Liked This, You're Sure to Love That," *New York Times*, November 23, 2008.

88 Thompson.

89 See Shinhyun Ahn and Chung-Kon Shi, "Exploring Movie Recommendation System Using Cultural Metadata," in *Transactions on Edutainment II*, ed. Z. Pan et al. (Berlin: Springer-Verlag, 2009), 119–134; and István Pilászy and Domonkos Tikk, "Recommending New Movies: Even a Few Ratings Are More Valuable than Metadata," *RecSys '09: Proceedings of the Third ACM Conference on Recommender Systems* (2009): 93–100.

90 John Lees-Miller et al., "Does Wikipedia Information Help Netflix Predictions?," *2008 Seventh International Conference on Machine Learning and Applications* (2008): 337–343.

91 Pilászy and Tikk, "Recommending New Movies."

92 Xavier Amatriain and Justin Basilico, "Netflix Recommendations: Beyond the 5 Stars (Part 1)," *Netflix Tech Blog*, April 5, 2012, http://techblog.netflix.com/2012 /04/netflix-recommendations-beyond-5-stars.html.

93 Finn, *What Algorithms Want*, 93.

94 Finn, 97.

95 Amatriain and Basilico, "Netflix Recommendations."

96 Christian Sandvig, "Corrupt Personalization," *Social Media Collective* (blog), June 26, 2014, https://socialmediacollective.org/2014/06/26/corrupt -personalization/.

97 Donna Bogatin, "Digg Democracy: Can Algorithms Trump Human Nature?," ZDNet, October 2, 2006, http://www.zdnet.com/blog/micro-markets/digg -democracy-can-algorithms-trump-human-nature/495; R. W. Gehl, "The Archive and the Processor: The Internal Logic of Web 2.0," *New Media and Society* 13, no. 8 (December 1, 2011): 1228–1244, https://doi.org/10.1177 /1461444811401735.

98 "Digg.com Site Info," Alexa, June 8, 2011, http://www.alexa.com/siteinfo/digg .com.

99 Samanth Subramanian, "Meet the Macedonian Teens Who Mastered Fake News and Corrupted the US Election," *Wired*, February 15, 2017, https://www.wired .com/2017/02/veles-macedonia-fake-news/.

100 James Surowiecki, *The Wisdom of Crowds* (New York, NY: Anchor, 2005).

101 Jodi Dean, *Publicity's Secret: How Technoculture Capitalizes on Democracy* (Ithaca, N.Y.: Cornell University Press, 2002), 3.

102 Dean, 4.

103 Axel Bruns, *Gatewatching: Collaborative Online News Production* (New York: Peter Lang International Academic, 2005), 1–2.

104 John Gapper, "The Digital Democracy's Emerging Elites," *Financial Times*, September 25, 2006, https://www.ft.com/content/b75779ae-4bf0-11db-90d2 -0000779e2340.

105 Matthew Hindman, *The Myth of Digital Democracy* (Princeton, N.J.: Princeton University Press, 2008), 5.

106 Gapper, "Digital Democracy's Emerging Elites."

107 Lee Rainie, John B. Horrigan, and Michael Cornfield, "The Internet and Campaign 2004," *Pew Research Center's Internet and American Life Project* (blog), March 6, 2005, http://www.pewinternet.org/2005/03/06/the-internet-and -campaign-2004/.

108 Ian Bogost, *Persuasive Games: The Expressive Power of Videogames* (Cambridge, Mass.: MIT Press, 2010).

109 "What Is Digg?," Digg, July 26, 2010, accessible on the Internet Archive's Wayback Machine, https://web.archive.org/web/20100726064954/http://about.digg.com/.

110 Tarleton Gillespie, "The Politics of 'Platforms,'" *New Media and Society* 12, no. 3 (May 1, 2010): 347–364, https://doi.org/10.1177/1461444809342738.

111 Kevin Rose, "Digg: New Algorithm Changes," *Digg the Blog,* January 23, 2008, https://web.archive.org/web/20081103073528/http://blog.digg.com:80/?p=106.

112 Muhammad Saleem, "It's the (Other) Algorithm, Stupid! Understanding DiggRank," *Search Engine Land* (blog), November 28, 2007, http://search engineland.com/it%E2%80%99s-the-other-algorithm-stupid-understanding -diggrank-12790.

113 Saleem.

114 Rand Fishkin, "Top 100 Digg Users Control 56% of Digg's HomePage Content," *Seomoz* (blog), October 10, 2007, https://moz.com/blog/top-100-digg-users -control-56-of-diggs-homepage-content.

115 Fishkin.

116 Richard Adhikari, "The Technological Tyranny of the Minority," TechNews-
World, September 10, 2010, http://www.technewsworld.com/story/70790.html.

117 Fishkin, "Top 100 Digg Users."

118 Muhammad Saleem, "The Power of Digg Top Users (One Year Later)," *ProNet
Advertising* (blog), October 11, 2007, http://www.pronetadvertising.com/articles
/the-power-of-digg-top-users-one-year-later34409.html.

119 "Digg's Failing Democracy," Download Squad, accessed July 6, 2013, http://
downloadsquad.switched.com/2007/10/22/diggs-failing-democracy/; Navneet
Alang, "Democracy Is Overrated: On Digg's Unlikely Rebirth," *Hazlitt* (blog),
May 8, 2013, http://www.randomhouse.ca/hazlitt/blog/democracy-overrated
-digg%E2%80%99s-unlikely-rebirth; Chris Wilson, "The Wisdom of the
Chaperones," *Slate*, February 22, 2008, http://www.slate.com/articles/technology
/technology/2008/02/the_wisdom_of_the_chaperones.html.

120 David Cohn, "Hunting Down Digg's Bury Brigade," *Wired*, March 1, 2007; "Digg
Patriots," *Schott's Vocab* (blog), *New York Times*, August 13, 2010, http://schott
.blogs.nytimes.com/2010/08/13/digg-patriots/; Om Malik, "A Very Angry Digg
Nation," GigaOM, May 1, 2007, http://gigaom.com/2007/05/01/a-very-angry
-digg-nation/.

121 Jacob Gower, "Digg Army: Right in Line," *ForeverGeek* (blog), April 19, 2006,
http://forevergeek.com/news/digg_corrupted_editors_playground_not
_userdriven_website.php.

122 Jodi Dean, *Democracy and Other Neoliberal Fantasies: Communicative Capitalism
and Left Politics* (Durham, N.C.: Duke University Press, 2009).

123 Ryan Block, "Users Turn against Digg: Anatomy of a Massive Online Revolt,"
Ryan Block (blog), May 1, 2007, http://ryanblock.com/2007/05/users-turn-against
-digg-anatomy-of-a-massive-online-revolt/.

124 Tamar Weinberg, "Social Media Revolt: Digg and Democracy," Search Engine
Roundtable, May 2, 2007, http://www.seroundtable.com/archives/013339.html.

125 Block, "Users Turn against Digg."

126 Kevin Rose, "Digg the Blog » Blog Archive » Digg This: 09-F9-11-02-9d-74-E3-5
b-D8-41-56-C5-63-56-88-C0," *Digg the Blog*, May 4, 2007, https://web.archive.org
/web/20070515231849/http://blog.digg.com/?p=74.

127 David Cohn, "Hunting Down."

128 Cohn.

129 P. Chandra, "New Digg Algorithm Favors Diversity for Front Page Stories,"
Quick Online Tips, January 25, 2008, http://www.quickonlinetips.com/archives
/2008/01/new-digg-algorithm-favors-diversity-for-front-page-stories/.

130 Rose, "Digg: New Algorithm Changes."

131 Gapper, "Digital Democracy's Emerging Elites."

132 Jordan Golson, "New Digg Algorithm Angers the Social Masses," Gawker,
January 23, 2008, http://gawker.com/348338/new-digg-algorithm-angers-the
-social-masses.

133 Tad Hogg and Kristina Lerman, "Social Dynamics of Digg," *EPJ Data Science* 1,
no. 5 (2012), https://doi.org/10.1140/epjds5.

134 Quoted in Golson, "New Digg Algorithm."

135 J. D. Rucker, "On Digg Today, Everything Went Better than Expected," *Soshable*
(blog), September 2, 2010, http://soshable.com/on-digg-today-everything-went
-better-than-expected/.

136 "Digg.com Site Info."

137 Charles Arthur, "Digg Loses a Third of Its Visitors in a Month: Is It Deadd?," *Guardian*, June 3, 2010, http://www.guardian.co.uk/technology/blog/2010/jun/03/digg-dead-falling-visitors.

138 TheKipper, comment on "Death of Digg Is Not Exaggerated" comment board, Digg, Oxt 2010, https://web.archive.org/web/20101007185352/http://digg.com:80/news/worldnews/death_of_digg_is_not_exaggerated_news.

139 Rucker, "On Digg Today."

140 Rucker.

141 Endersgame, comment on "Did Digg Game Its Own System to Benefit Publisher Partners?" comment board, Digg, October 26, 2010, https://web.archive.org/web/20101101094918/http://digg.com:80/news/technology/did_digg_game_its_own_system_to_benefit_publisher_partners.

142 RobertHillberg, "Important Development at Digg," October 26, 2010. https://web.archive.org/web/20101028160031/http://about.digg.com/blog/important-development-digg available.

143 Jared Keller, "Digg's Newest Corporate Sponsor: BP America," *Atlantic*, October 7, 2010.

144 Alison Diana, "Digg Lays Off 37% of Staff," *InformationWeek*, October 26, 2010, http://www.informationweek.com/software/soa-webservices/digg-lays-off-37-of-staff/228000040.

145 Chris Burns, "Digg Sale Splits the Company Three Ways for $16m Total," *Slashgear*, July 13, 2012, http://www.slashgear.com/digg-sale-splits-the-company-three-ways-for-16m-total-13238530/.

Chapter 4 Love's Labor's Logged

Epigraph: T. Jay Mathews, "Operation Match," *Harvard Crimson*, November 3, 1965, http://www.thecrimson.com/article/1965/11/3/operation-match-pif-you-stop-to/.

1 Over the last few years, more and more of these sites have relaxed their rules concerning whom users can contact on their sites and when they can do so, presumably to compete with the many free sites that have begun appearing.

2 Dawn Shepherd, *Building Relationships: Online Dating and the New Logics of Internet Culture* (Lanham, Md.: Rowman and Littlefield, 2016), 59.

3 Nick Paumgarten, "Looking for Someone," *New Yorker*, July 4, 2011, http://www.newyorker.com/reporting/2011/07/04/110704fa_fact_paumgarten.

4 Virginia Vitzthum, *I Love You, Let's Meet: Adventures in Online Dating* (New York: Little, Brown, 2007), 90; Yagan quoted in Paumgarten, "Looking for Someone," 12.

5 Eva Illouz, *Cold Intimacies: The Making of Emotional Capitalism* (Cambridge, U.K.: Polity, 2007), 79.

6 Lukas Brozovsky and Vaclav Petricek, "Recommender System for Online Dating Service," preprint, submitted March 9, 2007, 1, http://arxiv.org/abs/cs/0703042.

7 Alison P. Lenton, Barbara Fasolo, and Peter M. Todd, "'Shopping' for a Mate: Expected versus Experienced Preferences in Online Mate Choice," *IEEE Transactions on Professional Communication* 51, no. 2 (June 2008): 169–182, https://doi.org/10.1109/TPC.2008.2000342.

8 James Houran et al., "Do Online Matchmaking Tests Work? An Assessment of Preliminary Evidence for a Publicized 'Predictive Model of Marital Success,'" *North American Journal of Psychology* 6, no. 3 (2004): 507–526.

9 Caitlin Dewey, "Does Online Dating Work? Let's Be Honest: We Have No Idea," *Washington Post*, September 30, 2014, https://www.washingtonpost.com/news/the-intersect/wp/2014/09/30/does-online-dating-work-lets-be-honest-we-have-no-idea/.

10 Mark Hendrickson, "OKCupid Spins Off Quiz Engine, HelloQuizzy," *Washington Post*, June 18, 2008, http://www.washingtonpost.com/wp-dyn/content/article/2008/06/18/AR2008061801498.html.

11 Quoted in Vitzthum, *I Love You, Let's Meet*, 83.

12 Quoted in Vitzthum, 84.

13 Rufus Griscom, "Why Are Online Personals So Hot?," *Wired*, November 1, 2002, https://www.wired.com/2002/11/why-are-online-personals-so-hot/.

14 Aaron Smith and Monica Anderson, "5 Facts about Online Dating," *Pew Research Center FactTank* (blog), February 29, 2016, http://www.pewresearch.org/fact-tank/2016/02/29/5-facts-about-online-dating/.

15 Sherry Turkle, *Alone Together: Why We Expect More from Technology and Less from Each Other* (New York: Basic Books, 2011), 1.

16 Illouz, *Cold Intimacies*, 86.

17 David Sköld, "The Commercial Construction of a Perfect Date," *Pink Machine Papers* 24, no. 3 (2005): 3; Lenton, Fasolo, and Todd, "'Shopping' for a Mate"; Jeana H. Frost et al., "People Are Experience Goods: Improving Online Dating with Virtual Dates," *Journal of Interactive Marketing* 22, no. 1 (January 2008): 51–61, https://doi.org/10.1002/dir.20106.

18 Jenée Desmond-Harris, "Seeking My Race-Based Valentine Online," *Time*, February 22, 2010, http://www.time.com/time/magazine/article/0,9171,1963768,00.html.

19 These findings were corroborated by an earlier University of California, Irvine, study, which found that users of Yahoo! Personals were extremely biased in their racial preferences.

20 Christian Rudder, "Race and Attraction, 2009–2014," *OkCupid Blog*, September 10, 2014, https://theblog.okcupid.com/race-and-attraction-2009-2014-107dcbb4f060.

21 Eli J. Finkel et al., "Online Dating: A Critical Analysis from the Perspective of Psychological Science," *Psychological Science in the Public Interest* 13 (2012): 3.

22 Finkel et al., 3.

23 Lenton, Fasolo, and Todd, "'Shopping' for a Mate."

24 Eli J. Finkel and Susan Sprecher, "The Scientific Flaws of Online Dating Sites," *Scientific American*, May 8, 2012, https://www.scientificamerican.com/article/scientific-flaws-online-dating-sites/.

25 Paul Aditi, "Is Online Better than Offline for Meeting Partners? Depends: Are You Looking to Marry or to Date?," *Cyberpsychology, Behavior and Social Networking* 17, no. 10 (October 2014): 664–667, https://doi.org/10.1089/cyber.2014.0302; Jonathan D. D'Angelo and Catalina L. Toma, "There Are Plenty of Fish in the Sea: The Effects of Choice Overload and Reversibility on Online Daters' Satisfaction with Selected Partners," *Media Psychology* 20, no. 1 (January 2, 2017): 1–27, https://doi.org/10.1080/15213269.2015.1121827; Christiane Eichenberg, Jessica Huss, and Cornelia Küsel, "From Online Dating to Online

Divorce: An Overview of Couple and Family Relationships Shaped through Digital Media," *Contemporary Family Therapy* 39, no. 4 (December 1, 2017): 249–260, https://doi.org/10.1007/s10591-017-9434-x.

26 Dewey, "Does Online Dating Work?"

27 Quoted in Dan Slater, *Love in the Time of Algorithms: What Technology Does to Meeting and Mating* (London: Penguin Books, 2013), 93.

28 Karl Pearson, *The Life, Letters and Labours of Francis Galton: Volume 3—Part A: Correlation, Personal Identification and Eugenics* (Cambridge: Cambridge University Press, 1930), 31.

29 Francis Galton, *Hereditary Genius: An Inquiry into Its Laws and Consequences* (London: Macmillan, 1869).

30 Oliver P. John, Laura P. Naumann, and Christopher J. Soto, "Paradigm Shift to the Integrative Big Five Trait Taxonomy: History, Measurement, and Conceptual Issues," in *Handbook of Personality: Theory and Research*, 3rd ed., ed. Oliver P. John, Richard W. Robins, and Lawrence A. Pervin (New York: Guilford, 2010), 120.

31 John, Naumann, and Soto, 125.

32 Dan McAdams, "What Do We Know When We Know a Person?," *Journal of Personality* 63, no. 3 (1995): 374.

33 PerfectMatch.com also loosely based its matchmaking system on the Myers-Briggs test. Finkel et al., "Online Dating," 21.

34 Isabel Briggs Myers, *Gifts Differing: Understanding Different Personality Type*, with Peter B. Myers (Mountain View, Calif.: Consulting Psychologists Press, 1995), ix.

35 Myers, xiii.

36 Lucy Ash, "Can Personality Tests Identify the Real You?," BBC, July 6, 2012, http://www.bbc.co.uk/news/magazine-18723950.

37 Ash.

38 Myers, *Gifts Differing*, 129.

39 Quoted in Jennifer Hahn, "Love Machines," *AlterNet*, February 22, 2005, http://www.alternet.org/story/21291/love_machines/?page=3.

40 Nathan W. Hudson et al., "Coregulation in Romantic Partners' Attachment Styles: A Longitudinal Investigation," *Personality and Social Psychology Bulletin* 40, no. 7 (April 17, 2014): 845–857, https://doi.org/10.1177/0146167214528989.

41 Helen Fisher, *Why Him? Why Her? How to Find and Keep Lasting Love* (repr., New York: Holt Paperbacks, 2010).

42 "Helen Fisher's Personality Test," Anatomy of Love, accessed December 11, 2017, https://theanatomyoflove.com/relationship-quizzes/helen-fishers-personality-test/.

43 "About Us," Chemistry.com, accessed November 4, 2012, https://web.archive.org/web/20121118065114/http://www.chemistry.com/help/about.

44 Vitzthum, *I Love You, Let's Meet*, 82.

45 Diane Negra, *What a Girl Wants? Fantasizing the Reclamation of Self in Postfeminism* (London: Routledge, 2009), 12.

46 Anthea Taylor, *Single Women in Popular Culture: The Limits of Postfeminism* (New York: Palgrave Macmillan, 2011), 28.

47 Taylor, 3.

48 "Life Goes on a Date Arranged by Statistics," *Life*, August 3, 1942, 78.

49 "Life Goes on a Date," 79.

50 *Kiplinger* Washington editors, "Dating Services for Singles," *Kiplinger's Personal Finance*, May 1978.

51 "Life Goes on a Date," 79.

52 "Life Goes on a Date," 79.

53 "Life Goes on a Date," 79.

54 George W. Crane, "Tests for Husbands and Wives," Scribd, accessed November 4, 2012, http://www.scribd.com/doc/3086410/Tests-for-Husbands-and-Wives-.

55 Crane.

56 "Life Goes on a Date," 78.

57 *Kiplinger* Washington editors, "Dating Services for Singles," 23.

58 "New Rules for the Singles Game," *Life*, August 18, 1967, 61.

59 Marie Hicks, "Computer Love: Replicating Social Order through Early Computer Dating Systems," *Ada: A Journal of Gender, New Media, and Technology* 10 (2016), http://adanewmedia.org/2016/10/issue10-hicks/.

60 "New Rules," 65.

61 "New Rules," 65.

62 "How to Be Comfortable with Computer Dating," Compatibility computer dating advertisement, *Life*, August 8, 1969, NY1.

63 H. G. Cocks, *Classified: The Secret History of the Personal Column* (London: Cornerstone Digital, 2009), 161.

64 "The Secret History of the Personal Ad," BBC, February 5, 2009, http://news.bbc.co.uk/today/hi/today/newsid_7868000/7868125.stm.

65 Hicks.

66 Hicks.

67 Hicks.

68 John Hendel, "Old, Weird Tech: Computer Dating of the 1960s," *Atlantic*, February 14, 2011, http://www.theatlantic.com/technology/archive/2011/02/old-weird-tech-computer-dating-of-the-1960s/71217/.

69 Lucas Hilderbrand, "Undateable: Some Reflections on Online Dating and the Perversion of Time," *Flow*, April 22, 2011, http://flowtv.org/2011/04/undateable/.

70 Hilderbrand.

71 Hicks, "Computer Love."

72 Paumgarten, "Looking for Someone," 2.

73 Paumgarten, 2.

74 Paumgarten, 2.

75 Paumgarten, 2.

76 Quoted in Paumgarten, 2.

77 Mathews, "Operation Match."

78 Mathews.

79 Hendel, "Old, Weird Tech."

80 Hendel.

81 Hendel.

82 "London—Computer Marriages Aka Operation Match—Computer Matched Couples Get Together," film, 1968, 0:46, http://www.britishpathe.com/video/london-computer-marriages-aka-operation-match.

83 Eric Klien, "Electronic Matchmaker(Tm) Questionnaire—Draft 6.85," post to alt. personals Google group, December 26, 1992, https://groups.google.com/forum/#!msg/alt.personals/nrWxsw6_vdE/F89-qvvFy6gJ.

84 "Lycos to Buy Matchmaker.com for $44 Million," *Ad Age*, June 27, 2000, http://adage.com/article/news/lycos-buy-matchmaker-44-million/9696/.

85 eHarmony Staff, "Online Dating 101: Guided Communication," eHarmony Advice, October 1, 2007, https://web.archive.org/web/20160818085855 /http://www.eharmony.com/dating-advice/using-eharmony/online-dating-101 -guided-communication/.

86 Slater, *Love in the Time*, 92.

87 Smith and Anderson, "5 Facts about Online Dating."

88 "Dating Experts—Relationship and Dating Guidance on eHarmony," eHarmony, 2013, https://web.archive.org/web/20140110180316/http://www.eharmony.com /why/dating-experts/.

89 Janet Komblum, "eHarmony: Heart and Soul," *USA Today*, May 18, 2005, http://usatoday30.usatoday.com/life/people/2005-05-18-eharmony_x.htm.

90 Komblum.

91 Hahn, "Love Machines."

92 eHarmony, "eHarmony—A Social Revolution," June 3, 2013, video, 3:26, http://www.youtube.com/watch?v=vkfdghHbiwk&feature=youtube_gdata _player.

93 eHarmony, "eHarmony Ranks #1 for Most Online Marriages and Marital Satisfaction," press release, June 3, 2013, http://www.eharmony.com/press-release /57/#a.

94 Komblum, "eHarmony."

95 Carol Steffes, "Online Matchmaking," Family.org, 2002, accessible on the Internet Archive's Wayback Machine, http://web.archive.org/web /20030423181531/http://www.family.org/married/youngcouples/a0021741.cfm.

96 Steffes.

97 Komblum, "eHarmony."

98 John Cloud, "Is eHarmony Biased against Gays?," *Time*, June 1, 2007, http:// content.time.com/time/business/article/0,8599,1627585,00.html.

99 Steffes, "Online Matchmaking."

100 See Katherine Bindley, "eHarmony Duping Customers, Researchers Say," *Huffington Post*, last updated April 24, 2012, http://www.huffingtonpost.com /2012/04/23/eharmony-duping-customers-researchers-say_n_1446468.html; Amanda Lewis, "UCLA Professors Say eHarmony Is Unscientific and Its Customers Are 'Duped.' Here's Why," *Public Spectacle* (blog), *LA Weekly*, April 16, 2012, http://blogs.laweekly.com/arts/2012/04/eharmony_research _bullshit.php; and Finkel et al., "Online Dating."

101 Quoted in Bindley, "eHarmony Duping Customers."

102 Lori Gottlieb, "How Do I Love Thee?," *Atlantic*, March 2006, http://www .theatlantic.com/magazine/archive/2006/03/how-do-i-love-thee/304602/.

103 Gottlieb.

104 "Compatible Partners—Gay Dating and Lesbian Dating for the Relationship-Minded Gay Singles & Lesbian Singles," Compatible Partners, 2011, http://www .compatiblepartners.net/; Rachel Gordon, "eHarmony Settles Lawsuit over Gay Matchmaking," *SFGate*, January 27, 2010, http://www.sfgate.com/bayarea/article /EHarmony-settles-lawsuit-over-gay-matchmaking-3274826.php#src=fb.

105 "Compatible Partners Review," Cupid's Library, 2012, https://web.archive.org /web/20120918132033/http://www.cupidslibrary.com:80/sites/compatible -partners.

106 Michelle Garcia, "The Online Dating Game," *Advocate*, March 31, 2010, http://
www.advocate.com/politics/commentary/2010/03/31/online-dating-game.

107 Quoted in Randy Dotinga, "Online Dating Sites Aren't Holding People's Hearts,"
Christian Science Monitor, January 27, 2005, http://www.csmonitor.com/2005
/0127/p11s02-stin.html.

108 "eHarmony—First Step at Dating 101?," Examiner.com, accessed January 1, 2013,
http://www.examiner.com/article/eharmony-first-step-at-dating-101.

109 Quoted in John Tierney, "Hitting It Off, Thanks to Algorithms of Love," *New
York Times*, January 29, 2008, http://www.nytimes.com/2008/01/29/science
/29tier.html.

110 "CentACS WorkPlace and SchooPlace Big Five Profile," Center for Applied
Cognitive Studies, 2013, https://web.archive.org/web/20130402062706
/http://www.centacs.com/.; John, Naumann, and Soto, "Paradigm Shift," 125.

111 A. T. Fiore and J. S. Donath, "Homophily in Online Dating: When Do You Like
Someone like Yourself?," *CHI '05 Extended Abstracts on Human Factors in
Computing Systems* (2005): 1371, http://dl.acm.org/citation.cfm?id=1056919.

112 Hahn, "Love Machines."

113 Quoted in Hahn.

114 Illouz, *Cold Intimacies*, 78.

115 Illouz, 32.

116 Illouz, 76.

117 Michel Foucault, *Technologies of the Self: A Seminar with Michel Foucault*
(Amherst: University of Massachusetts Press, 1988), 17.

118 Foucault, 17.

119 Foucault, 18.

120 Foucault, 26.

121 In chapter 1, I discuss how this process leads to what Jean-Paul Sartre referred to
as "acting in bad faith."

122 Paumgarten, "Looking for Someone."

123 Paumgarten.

124 Illouz, *Cold Intimacies*, 112.

125 Illouz, 80.

126 Illouz, 79.

127 eHarmony's actual rejection page cannot be linked to, but a link to a screenshot of
how the page looked in 2010 can be found in Kim Noble, "Dating Agencies are
Anti-Ginger," *Kim Noble* (blog), accessed September 26, 2018, http://mrkimnoble
.com/kim-noble-on-dating-agencies-are-anti-ginger/.

128 Paul Farhi, "They Met Online, but Definitely Didn't Click," *Washington Post*,
May 13, 2007, http://www.washingtonpost.com/wp-dyn/content/article/2007/05
/12/AR2007051201350.html?hpid=topnews.

129 Steffes, "Online Matchmaking."

130 Farhi, "They Met Online"; Steffes, "Online Matchmaking."

131 Quoted in Komblum, "eHarmony."

132 "eHarmony—Cannot Provide Service."

133 While this practice is legal in online dating, Kroger was recently sued for
incorporating a similar test in its employee screening process under the Americans
with Disabilities Act.

134 "5 Worst Websites," *Time*, July 9, 2007, http://content.time.com/time/specials
/2007/article/0,28804,1638344_1638341,00.html; Cloud, "Is eHarmony Biased

against Gays?"; Farhi, "They Met Online"; Lisa Miller, "An Algorithm for Mr. Right," *Newsweek*, April 26, 2008, https://web.archive.org/web /20120208010259/http://www.thedailybeast.com/newsweek/2008/04/26/an -algorithm-for-mr-right.html.

135 Linda Hutcheon, *The Politics of Postmodernism* (New York: Routledge, 1989), 206.

136 Razmig, "Whites Only Dating Website—White People Meet," August 26, 2012, video, 1:13, https://www.youtube.com/watch?v=ysGsihiGQKY.

137 Comment board for Razmig.

138 Devin, August 27, 2012, comment on Razmig, "WhitesOnlyDating.Com . . . ," *RAZMIG* (blog), August 26, 2012, http://razmig.net/2012/08/26 /whitesonlydating-com/.

139 Kathy D, July 4, 2016, comment on Razmig.

140 Ben Jaffe, interview by the author, email, November 14, 2017.

141 Jaffe.

142 Quoted in "The Age of the Algorithm," produced by Delaney Hall, *99% Invisible* (blog), September 5, 2017, https://99percentinvisible.org/episode/the-age-of-the -algorithm/.

Chapter 5   The Mirror Phased

1 "Facial Beauty Analysis—Score Your Face," Anaface, accessed September 4, 2018, http://anaface.com/.

2 Brian Elgin, "New Anaface Facial Beauty Analysis Software Calculates Looks Instantly," PRWeb, May 13, 2009, http://www.prweb.com/releases/2009/05 /prweb2415524.htm.

3 Elgin.

4 "Facial Beauty Analysis."

5 Bernadette Wegenstein, *The Cosmetic Gaze: Body Modification and the Construction of Beauty* (Cambridge, Mass.: MIT Press, 2012), 109.

6 Alexandra Howson, *The Body in Society: An Introduction* (Cambridge, U.K.: Blackwell, 2004), 96.

7 Anne Marie Balsamo, *Technologies of the Gendered Body: Reading Cyborg Women* (Durham, N.C.: Duke University Press, 1996), 58.

8 Anthony Giddens, *Modernity and Self-Identity: Self and Society in the Late Modern Age* (Stanford, Calif.: Stanford University Press, 1991), 8.

9 L. Pecot-Hebert and H. Hennink-Kaminski, "I Did It for Me: Negotiating Identity and Agency," *Health, Culture and Society* 3, no. 1 (August 14, 2012): 84.

10 Kathryn Pauly Morgan, "Women and the Knife: Cosmetic Surgery and the Colonization of Women's Bodies," *Hypatia* 6, no. 3 (October 1, 1991): 36.

11 Amy Alkon, "The Truth about Beauty," *Psychology Today*, November 1, 2010, http://www.psychologytoday.com/articles/201011/the-truth-about-beauty.

12 Allan Sekula discusses how a similar standardization of body photography disciplined the bodies of criminals. See Allan Sekula, "The Body and the Archive," *October* 39 (1986): 3.

13 Robert Kotler, "How Cosmetic Surgeons Use Photos," WebMD, May 2012, video, 2:19, http://www.webmd.com/beauty/video/before-after-photos-cosmetic -surgery.

14 Gerald D. Nelson and John L. Krause, "Introduction," in *Clinical Photography in Plastic Surgery*, ed. Gerald D. Nelson and John L. Krause (Boston: Little, Brown, 1988), ix.

15 Harvey A. Zarem, "Standards of Medical Photography," in Nelson and Krause, *Clinical Photography in Plastic Surgery*, 73.

16 Nelson and Krause, ix.

17 Nelson and Krause, ix.

18 Jayne O'Donnell, "Cosmetic Surgery Laws Often Aren't Enough," *USA Today*, December 10, 2012, http://www.usatoday.com/story/money/2012/12/10/cosmetic-surgery-laws-effect-debated/1759839/.

19 O'Donnell.

20 Denis Campbell, "More Patients Sue Plastic Surgeons over Faulty Cosmetic Operations," *Guardian*, January 8, 2012, http://www.theguardian.com/society/2012/jan/08/patients-sue-plastic-surgeons-faulty.

21 Noah Scheinfeld, "Photographic Images, Digital Imaging, Dermatology, and the Law," *Archives of Dermatology* 140 (April 2004): 473.

22 Nelson and Krause, ix.

23 See Grant S. Hamilton III, "Photoshop Tips and Tricks Every Facial Plastic Surgeon Should Know," *Facial Plastic Surgery Clinics of North America* 18, no. 2 (May 2010): 283–328.

24 See Reha Yavuzer, Stefani Smirnes, and Ian Jackson, "Guidelines for Standard Photography in Plastic Surgery," *Annals of Plastic Surgery* 46, no. 3 (March 2001): 293–300; John DiSaia, Jeffrey Ptak, and Bruce Achauer, "Digital Photography for the Plastic Surgeon," *Plastic and Reconstructive Surgery* 102, no. 2 (August 1998): 569–73; and Joseph Niamtu III, "Image Is Everything: Pearls and Pitfalls of Digital Photography and PowerPoint Presentations for the Cosmetic Surgeon," *Dermatological Surgery* 30 (2004): 81–91.

25 "Digital Technology Gives New Jersey Plastic Surgery Patients a 'Sneak Peek' before Going under the Knife," PRWeb, March 7, 2009, http://www.prweb.com/releases/cosmetic/surgery/prweb2214634.htm.

26 "3D Medical Imaging Technology Software Company," Crisalix, 2012, https://web.archive.org/web/20120819080701/http://www.crisalix.com:80/en/about.

27 "3D Breast Implant Simulation Online: Web-Based Software | e-Stetix," Crisalix, 2012, https://web.archive.org/web/20120228185809/http://www.crisalix.com/en/products.

28 "3D Simulation Software for Plastic Surgeons: Testimonials," Crisalix, 2012, https://web.archive.org/web/20160328132141/http://www.crisalix.com/en/testimonials.

29 "3D Simulation Software."

30 "Canfield Scientific Inc," 2012, https://web.archive.org/web/20120216221938/http://www.canfieldsci.com:80/imaging_systems/imaging_software/Sculptor_3D_Simulation.html.

31 cssmedical, "Canfield Vectra 3D Imaging System," Plastic Surgery Center of Fairfield, February 20, 2011, video, 2:57, https://www.youtube.com/watch?v=ZUvb7dvFTMQ.

32 "Virtual Plastic Surgery | 3D Computer Imaging," Simoni Plastic Surgery, accessed September 4, 2018, http://www.drsimoni.com/virtual-plastic-surgery.htm.

33 "Virtual Plastic Surgery."

34 For one example, see "Ethnic Rhinoplasty in Los Angeles," 2012, http://www
.rodeodriverhinoplasty.com/ethnic-rhinoplasty.html.

35 "At Home Virtual Cosmetic Plastic Surgery Consultation Online via Skype,"
2012, https://web.archive.org/web/20120530224323/http://www
.rodeodriverhinoplasty.com:80/ethnic-rhinoplasty.html/.

36 Natasha Singer, "Is the 'Mom Job' Really Necessary?," *New York Times*, October 4, 2007, http://www.nytimes.com/2007/10/04/fashion/04skin.html.

37 Singer.

38 Susan Bordo, *Twilight Zones: The Hidden Life of Cultural Images from Plato to
O.J.* (Berkeley: University of California Press, 1997), 55.

39 Bordo, 50.

40 Bordo, 56.

41 Danya Hoenig, "'Virtual Plastic Surgery'—BeautySurge.com Announces Launch
of Cosmetic Surgery Digital Imaging Services," PRWeb, 2004, http://www.prweb
.com/releases/2004/03/prweb109961.htm.

42 Heather Hurst, "Digital Plastic Surgery," accessed September 4, 2018, http://www
.finishmyfoto.com/digital-plastic-surgery.html.

43 Steve Patterson, "Photoshop Tutorials: Easy Digital Nose Job in Photoshop,"
Photoshop Essentials, accessed September 4, 2018, http://www
.photoshopessentials.com/photo-editing/nose/.

44 Patterson.

45 compu-smart, "Models & Celebrities without Makeup: Digital Cosmetic Surgery
Facts," 2012, https://web.archive.org/web/20120522203022/http://compu-smart
.hubpages.com/hub/digital-cosmetic-surgery.

46 Bordo.

47 FaceTouchUp homepage, accessed September 4, 2018, http://www.facetouchup
.com/.

48 FaceTouchUp.

49 "Plastic Surgery IPhone App Development," 2011, https://web.archive.org/web
/20120815072445/http://www.facetouchup.com/products/plastic-surgery-iphone
-app-development.html.

50 In *The Body and Society*, Alexandra Howson describes a variety of ways in which
the commodification of the body (through forms of modification ranging from
jogging to tattoos) can become a method for identifying with a particular lifestyle.
See Howson, *Body in Society*.

51 "3D Simulation for Breast Augmentation Plastic Surgery," Sublimma, 2012,
https://web.archive.org/web/20120425211919/http://www.sublimma.com:80/.

52 "3D Simulation for Breast Augmentation."

53 "Online Plastic Surgery Simulator, Photo Editing Tool SurgeryMorph Helps
Patients Visualize Potential Outcomes," *Cision PRWeb*, December 1, 2010,
https://www.prweb.com/releases/plastic-surgery/photo-simulator/prweb4845314
.htm.

54 SurgeryMorph homepage, accessed September 5, 2018, http://www.surgerymorph
.com.

55 "SurgeryMorph Video Demonstration," SurgeryMorph, 2012, FLV file, https://
web.archive.org/web/20150222120748/http://www.surgerymorph.com:80/video
.php.

56 "About ModiFace," ModiFace, 2012, https://web.archive.org/web
/20120716031608/http://modiface.com:80/index.php?about.

57 "#1 Creator of Mobile/Tablet Beauty and Fashion Apps," ModiFace, 2012, https://web.archive.org/web/20120828122742/http://modiface.com:80/index.php?mobile.

58 "LiftMagic—Cosmetic Surgery and Anti-aging Makeover Tool," LiftMagic, accessed September 4, 2018, http://makeovr.com/liftmagic/liftupload.php.

59 Bal Harbour Plastic Surgery Associates, "Virtual Plastic Surgery Segment," September 15, 2010, video, 2:37, http://www.youtube.com/watch?v=rdooNK9zqAE.

60 "Heidi Yourself," MTV UK, 2010, https://web.archive.org/web/20120518065220/http://www.mtv.co.uk/heidiyourself.

61 Rosemina Nazarali, "Heidi Montag 'Heidi Yourself' Plastic Surgery App," 29 Secrets, 2011, http://29secrets.com/sections/daily-whisper/heidi-montag-heidi-yourself-plastic-surgery-app.

62 Virtual Plastic Surgery Simulator homepage, accessed September 4, 2018, http://www.plastic-surgery-simulator.com/.

63 Hot or Not homepage, accessed September 4, 2018, http://hotornot.com.

64 Sarah Kershaw, "The Sum of Your Facial Parts," New York Times, October 8, 2008.

65 Kershaw.

66 Kershaw.

67 Tommer Leyvand et al., "Data-Driven Enhancement of Facial Attractiveness," ACM Transactions on Graphics (proceedings of ACM SIGGRAPH 2008, Los Angeles, 2008) 27, no. 3 (2008): 2, http://leyvand.com/beautification2008/attractiveness2008.pdf.

68 Kershaw, "Sum of Your Facial Parts."

69 Leyvand et al., "Data-Driven Enhancement," 1.

70 Maggie Bullock, "Tech Support," Elle, April 14, 2008, http://www.elle.com/beauty/golden-ratio-perfect-face.

71 Leyvand et al., "Data-Driven Enhancement," 1.

72 Sam Levin, "A Beauty Contest Was Judged by AI and the Robots Didn't Like Dark Skin," Guardian, September 8, 2016, http://www.theguardian.com/technology/2016/sep/08/artificial-intelligence-beauty-contest-doesnt-like-black-people.

73 Levin.

74 Levin.

75 Kershaw, "Sum of Your Facial Parts"; Leyvand et al., "Data-Driven Enhancement," 8.

76 Amit Kagian et al., "A Machine Learning Predictor of Facial Attractiveness Revealing Human-like Psychophysical Biases," Vision Research 48, no. 2 (January 2008): 2.

77 Francis Galton, "Composite Portraits," Journal of the Anthropological Institute of Great Britain and Ireland 8 (1878): 136.

78 Gillian Rhodes and Leslie Zebrowitz, eds., Facial Attractiveness: Evolutionary, Cognitive, and Social Perspectives, Advances in Visual Cognition 1 (Westport, Conn.: Ablex, 2001); Adam J. Rubenstein, Judith H. Langlois, and Lori A. Roggman, "What Makes a Face Attractive and Why: The Role of Averageness in Defining Facial Beauty," in Rhodes and Zebrowitz, Facial Attractiveness, 1–33; Tim Valentine, Stephen Darling, and Mary Donnelly, "Why Are Average Faces Attractive? The Effect of View and Averageness on the Attractiveness of Female

Faces," *Psychonomic Bulletin and Review* 11, no. 3 (June 1, 2004): 482–487, https://doi.org/10.3758/BF03196599.

79 Kershaw, "Sum of Your Facial Parts."

80 Leslie Zebrowitz and Gillian Rhodes, introduction to *Facial Attractiveness*, ed. Zebrowitz and Rhodes, vii.

81 Kershaw, "Sum of Your Facial Parts."

82 Bullock, "Tech Support."

83 "'Beauty Micrometer' Analyzes Facial Flaws for Makeup (Jan, 1935)," Modern Mechanix, April 15, 2006, http://blog.modernmechanix.com/beauty-micrometer -analyzes-facial-flaws-for-makeup/.

84 "Making Beauty," MBA California, accessed September 25, 2013, https://web .archive.org/web/20120511155646/http://www.beautyanalysis.com:80/index2 _mba.htm.

85 Stephen Marquardt and Eugene Gottlieb, "Dr. Stephen R. Marquardt on the Golden Decagon and Human Facial Beauty," *Journal of Clinical Oncology* 36, no. 6 (2002): 342.

86 "Beauty Assessment," Face Beauty Rank, accessed September 4, 2018, http://www .facebeautyrank.com/beautyassessment.html.

87 Face Beauty Rank homepage, accessed September 4, 2018, http://www .facebeautyrank.com/.

88 Face Beauty Rank homepage.

89 Face Beauty Rank homepage.

90 Amanda Parker, "Beauty Analyzer," App Store, June 4, 2012, https://itunes.apple .com/us/app/beauty-analyzer/id471731380?mt=8.

91 Kendra Schmid, David Marx, and Ashok Samal, "Computation of a Face Attractiveness Index Based on Neoclassical Canons, Symmetry, and Golden Ratios," *Pattern Recognition* 41, no. 8 (2007): 2710.

92 Schmid, Marx, and Samal, 2716.

93 Murray Bourne, "The Math behind the Beauty," Interactive Mathematics, last modified April 1, 2018, http://www.intmath.com/numbers/math-of-beauty.php.

94 Ana C. Andres del Valle and Agata Opalach, "The Persuasive Mirror," unpublished manuscript (Sophia Antipolis, France: Accenture Technology Labs, n.d.).

95 "What Is Captology?," Stanford Persuasive Tech Lab, accessed September 4, 2018, http://captology.stanford.edu/about/what-is-capatology.html.

96 B. J. Fogg, "Thoughts on Persuasive Technology," November 2010, Stanford Persuasive Tech Lab, http://captology.stanford.edu/resources/thoughts-on -persuasive-technology.html.

Conclusion

1 Sara McTigue, "It's All about Choices: Getting Your Toddler to Cooperate," EverydayFamily, June 14, 2013, http://www.everydayfamily.com/blog/its-all -about-choices-getting-your-toddler-to-cooperate/.

2 McTigue.

3 KaelinRae, June 14, 2013, comment on McTigue, "It's All about Choices.

4 For a more detailed discussion of this process, see Toby Miller, *Blow Up the Humanities* (Philadelphia: Temple University Press, 2012).

5 "Our Latest," Google, accessed September 5, 2018, https://www.google.com/about/.

Index

agency, personal. *See* personal agency

Agents (internet start-up), 73, 75, 77–78, 79
    See also Maes, Pattie

agents. *See* autonomous agents

AI (artificial intelligence)
    collaborative learning and, 60
    conceptions of choice and, 29, 40
    Eliza (Rogerian therapy program), 39
    Maes's work with, 60–61
    racially biased training of, 178–179
    Turing's ideas on gender and, 36, 37–38
    See also autonomous agents

akrasia, 70

algorithmic culture
    defining, 12–13
    human bias in, 9, 15–16, 41, 131
    phallogocentrism in, 88–90, 99, 102, 109, 113–114
    subverting, 20, 24, 30, 52, 103–104, 110–111, 116, 187

algorithms
    collecting user data for, 5, 19, 20–21, 39, 109
    complexity of, 13–15, 20, 84, 87, 93, 103, 110
    corporate representation of, 8, 11–12, 48, 85–86, 110, 126
    examples of racism in, 14–15, 35, 130–132, 154–155
    as imposing identity, 86–87, 94–95, 97, 157, 161–162
    limitations of, 20–22, 34–35, 93–94, 99, 187

the myth of "benign", 3, 10–11, 14–15, 19, 33, 39, 42

"objectivity" of, 11, 13, 83–84, 156–157, 160, 180

personal agency in, 15–16, 77, 86–87, 89, 103–105, 111–112, 124, 152–155

reading strategies for, 93–94

See also collaborative filtering; recommendation systems

Amazon
    use of collaborative filtering by, 6, 18–19, 34–35
    racism in Prime delivery services, 14–15

Anaface (cosmetic surgery website), 25, 158–159

artificial intelligence. *See* AI

automated computer technologies.
    See recommendation technologies

automated recommendations. *See* recommendations

autonomous agents, 59–60, 63, 65–66, 69, 76–77

beauty analyzers, 177–179, 181–183

Beauvoir, Simone de, 36

"big data", 3–4, 75

bourgeoisie (white American)
    as "ideal" consumers, 2, 33, 59
    recommendation systems and, 3–4, 7, 21, 64, 131
    See also capitalism

About the Author

JONATHAN COHN is an assistant professor at the University of Alberta.